KUHN'S *THE STRUCTURE OF SCIENTIFIC REVOLUTIONS*

Continuum *Reader's Guides*

Continuum's *Reader's Guides* are clear, concise and accessible introductions to classic works of philosophy. Each book explores the major themes, historical and philosophical context and key passages of a major philosophical text, guiding the reader toward a thorough understanding of often demanding material. Ideal for undergraduate students, the guides provide an essential resource for anyone who needs to get to grips with a philosophical text.

KUHN'S *THE STRUCTURE OF SCIENTIFIC REVOLUTIONS*

A Reader's Guide

JOHN PRESTON

continuum

Continuum International Publishing Group
The Tower Building   80 Maiden Lane
11 York Road   Suite 704
London SE1 7NX   New York, NY 10038
www.continuumbooks.com

British Library Cataloguing-in-Publication Data
A catalogue record for this book is available from the British Library.

ISBN-10: HB: 0-8264-9375-0
        PB: 0-8264-9376-9
ISBN-13: HB: 978-0-8264-9375-0
        PB: 978-0-8264-9376-7

Library of Congress Cataloging-in-Publication Data
Preston, John.
Kuhn's "The structure of scientific revolutions": a reader's guide /
John Preston.
p. cm.
Includes bibliographical references and index.
ISBN 978-0-8264-9375-0 – ISBN 978-0-8264-9376-7
1. Kuhn, Thomas S. Structure of scientific revolutions.   2. Science–
Philosophy.   3. Science–History.   I. Kuhn, Thomas S. Structure of
scientific revolutions.   II. Title.
Q175.K953P74   2008
501–dc22   2007044434

Typeset by Servis Filmsetting Ltd, Manchester
Printed and bound in Great Britain by
MPG Books Ltd, Bodmin, Cornwall

CONTENTS

PREFACE

Thomas Kuhn's *The Structure of Scientific Revolutions* (henceforth 'SSR') was not written merely for an academic readership, and isn't a difficult book to read. But it has certainly proved a difficult and immensely controversial book to *interpret*, and this is where this *Guide* is intended to help.

SSR is one of the two or three most important contributions to twentieth-century history and philosophy of science, one of the most recent philosophical 'classics', works that serious students of philosophy (not just the philosophy of science), and many serious students of several other subjects, have to get to grips with. I suspect it is still likely to be read in a hundred years' time.

As an introductory book aimed at students just encountering the philosophy of science, this *Guide* focuses almost exclusively on the general picture of science presented in SSR, rather than on the book's historical material, which of course would have to be considered in detail in any more thorough account. My main aim has been to raise probing questions about SSR (at least partly within the 'discussion points' appended to each section of the *Guide*). I have not tried to cover Kuhn's two historical books, or to give more than pointers to the content of his later work, even though most of this consisted of elaborations of the themes of SSR.

I should like to thank my colleagues in the Department of Philosophy at the University of Reading, and especially my former colleague Hanjo Glock (now at the University of Zürich). My reading of Kuhn draws on widely differing thoughts gleaned from the work of Hanjo, Peter Hacker, Robert Arrington, Ian Hacking, Alasdair MacIntyre and Brendan Larvor. I would also like to thank Stanley Cavell for very helpful correspondence concerning Kuhn and Wittgenstein, and Stefano Gattei, for reading the entire book in manuscript and making very helpful suggestions before it went to press. (None of the above should be saddled with my reading or my views, though.)

My main thanks, though, go to my wonderful wife, Roberta, the love of my life.

KUHN'S BOOKS – ABBREVIATIONS

In this *Guide*, Kuhn's books are referred to as follows:

CR: *The Copernican Revolution: Planetary Astronomy in the Development of Western Thought* (Cambridge, MA: Harvard University Press, 1957).

SSR: *The Structure of Scientific Revolutions*, 3rd edition (Chicago: University of Chicago Press, 1996).

ET: *The Essential Tension: Selected Studies in Scientific Tradition and Change* (Chicago: University of Chicago Press, 1977).

RSS: *The Road Since* Structure: *Philosophical Essays, 1970–1993*, eds J. B. Conant and J. Haugeland (Chicago: University of Chicago Press, 2000).

All page references without a reference to some specific text are to SSR.

CHAPTER 1

CONTEXT

KUHN'S LIFE AND WORK

Thomas Samuel Kuhn was born in Cincinnati, Ohio, in 1922, into a non-practising Jewish family. He attended progressive schools in New York City, New York State and Pennsylvania, before his parents moved him to a school in Connecticut, in preparation to enter Harvard University. Having done well at this school, especially in maths and science subjects, he went on to Harvard in 1940, intending to major in physics. Although concentrating on electromagnetism and electronics during his undergraduate studies, he also took some history, and courses on classical and modern philosophy. The Second World War disrupted his curriculum, but he graduated in only three years. By 1943 he was already thinking that his future career might lie in philosophy. He went to work for the Radio Research Laboratory of the US Office of Scientific Research and Development, first at Harvard, then conducting research on radar countermeasures with bomber groups in England, and finally examining radar installations on the Continent as the German forces retreated.

After the war ended in 1945, although Kuhn enrolled as a graduate student in theoretical physics at Harvard, he soon expressed doubts about his intended course of study, and received permission to take courses in other disciplines also, including philosophy. He studied with the logician Henry M. Sheffer, and came under the influence of the epistemology, logic and semantics of the pragmatist philosopher C. I. Lewis. This period confirmed his interest in the subject, and around this time he became convinced that a book like SSR needed to be written (ET, p. x). However, he decided to finish his doctorate in physics before changing disciplines.

James Bryant Conant, the President of Harvard, was at that time setting up a 'General education in science' programme, designed to inform undergraduate students of non-science subjects about the nature of science. Kuhn, having made himself known to the faculty for editing the student newspaper and being prepared to write reports on matters affecting students, was invited by Conant to become an assistant on a history of science course within this programme. (History of science was only just finding its feet as an academic discipline, and there were very few people employed to teach it in the whole US.) The case study on the history of mechanics which Conant asked Kuhn to do helped further shift his interest from physics to the history of science (although that was, for Kuhn, the means to a *philosophical* end).

It was during the preparation for this course, in 1947, when Kuhn was reading about Aristotle's physics, that he had an epiphany or revelation about the way to do history of science. Thinking from the point of view of more recent physical science, he found, produced the puzzling consequence that Aristotle, a supremely talented observer of many of nature's aspects, seemed instead to have made blatant errors when it came to physics. How could this have been so, and how then could his physics have been taken seriously for so long? A reading of the history of science which gave rise to such perplexities, Kuhn felt, had to have gone wrong.[1] But '[o]ne memorable (and very hot) summer day those perplexities suddenly vanished. I all at once perceived the connected rudiments of an alternate way of reading the texts with which I had been struggling' (ET, p. xi). By learning more about the universe as Aristotle conceived it, Kuhn came to see Aristotle's physics in a different way, a way which was more like Aristotle's than like the world-view of modern physics. Via a process that he later likened to a Gestalt-switch he had, as it were, got inside Aristotle's head (RSS, pp. 276, 280, 293, 315). In calling SSR 'a project originally conceived almost fifteen years ago' (p. v, see also RSS, p. 292), Kuhn explicitly traced the book back to this 1947 epiphany.

This kind of conceptual readjustment on the part of the *historian*, Kuhn felt, mirrored that which must have occurred to physicists themselves during the history of that discipline. There must have been a global kind of conceptual change within physics, a change that wasn't merely the addition of new material to what was already known, or the correction of mistakes, but rather what the historian

Herbert Butterfield had described as 'putting on a different kind of thinking-cap'. It was in 1947, then, that Kuhn 'stumbled upon' the concept that would come to mean so much to him, that of a *scientific revolution* (ET, p. xvi).

Kuhn had already 'read and admired a good deal' (RSS, p. 285) the well-known book by the philosophically-trained historian Arthur Lovejoy, *The Great Chain of Being* (Lovejoy 1936), one of the founding texts of the discipline known as the history of ideas. At the time of his Aristotle epiphany, though, Kuhn's colleague, the Harvard historian of science I. B. Cohen, suggested to him that he read the *Études galiléennes* (*Galileo Studies*) by the émigré Russian historian of philosophy and science Alexandre Koyré (1892–1964). Although this book had been published in 1939, it became known only after the war. Koyré's work, and the '*historiographic revolution*' it initiated exerted a strong influence on Kuhn, as we shall see. Koyré showed Kuhn that what Lovejoy did with the history of ideas could also be done with the history of *science* (RSS, p. 285).

Although its originator was probably Auguste Comte, Koyré had a central role in establishing and clarifying the concept of *the scientific revolution*. 'The scientific revolution of the seventeenth century', he wrote, was 'one of the most important, perhaps even *the* most important, [mutation in human thought] since the invention of the Cosmos by Greek thought' (Koyré 1978, p. 1). The idea of a single such revolution was then taken up by Butterfield, A. Rupert Hall and other historians, who located it variously within the period 1300–1800, but usually in the sixteenth and seventeenth centuries. Butterfield claimed that this revolution 'overturned the authority in science not only of the middle ages but of the ancient world', and that it was important enough to constitute 'the real origin both of the modern world and of the modern mentality' (Butterfield 1949, p. viii). Koyré had already used the plural term 'revolution*s* [of scientific ideas]' (Koyré ibid.), but it was Kuhn who explicitly took the concept of *the* scientific revolution and used it in what he calls an 'extended conception' (pp. 7, 8), characterizing *several* different episodes in the history of science as 'scientific revolutions'.

With Conant's sponsorship, Kuhn was elected a Junior Fellow in Harvard's Society of Fellows in 1948. This was his first opportunity not only to educate himself deeply in the history of science, the subject he had applied to do, but also to read in related areas, such as sociology, the history of ideas, psychology, anthropology and linguistics, and it was also when he encountered the ideas of the Harvard philosopher W. v. O. Quine. Kuhn drew upon all these sources in constructing SSR, but it resulted mainly from his sustained immersion in past science during the 1950s. This began to undermine the basic conceptions he had drawn from his own scientific training and his interest in the philosophy of science (p. v).

Kuhn submitted his dissertation, which concerned a new way of measuring the cohesive energy of certain types of metals, and was awarded a doctorate in physics in 1949. His first publications, in 1950 and 1951, were in physics and applied mathematics, but by the time they appeared he had, as he later put it, 'abandoned science for its history' (ET, p. x).

When the well-known philosopher of science Karl Popper gave his William James lectures at Harvard early in 1950, Kuhn made his acquaintance.[2] Popper pointed Kuhn towards the work of the Paris-based philosopher Émile Meyerson (1859–1933), whose approach to history Kuhn came to admire. Meyerson's philosophy of science, which had Koyré's support, was strongly opposed to the then-popular approach known as *positivism*, but his tendency to see the rational structure of human thought (in the form of a small group of scientific conservation principles) operating throughout natural science caused him to overlook the issue of scientific *change*, which was Kuhn's concern.

Kuhn travelled to Europe in the summer of 1950, meeting first with philosophers and historians of science at University College London, where there was a history of science programme, and at Oxford, and then travelling to France. Koyré, who was based in Paris, but who Kuhn had already met when giving lectures in the US, had provided him with a letter of introduction to the French philosopher Gaston Bachelard (1884–1962), one of whose doctoral dissertations Kuhn had read (Bachelard 1927). While their meeting was apparently unproductive (RSS, pp. 284–5), Bachelard's concepts of 'epistemological break' (*coupure épistémologique*) and 'mutation', transmitted via Koyré, are relatives of and may even have been an inspiration behind Kuhn's idea of scientific revolutions.

During this trip Kuhn also encountered the work of historians of science such as Hélène Metzger and Anneliese Maier, which he came to admire. His own writings were the route by which the work of these continental European historians and philosophers of science somewhat surreptitiously entered the Anglo-American scene. Kuhn's debt to these figures, whose work wasn't well known in the US and Britain, is undoubtedly one of the main reasons why SSR seemed so new and created such a fuss there. But the idea that became most familiar from his work, of discontinuities in the history of science, was already commonplace in France.

The year after he returned from Europe Kuhn delivered the 1951 Lowell lectures at Harvard on 'The Quest for Physical Theory', and began publishing in the main history of science journal, *Isis*. The Lowell lectures anticipate some of the central features and concepts of SSR. Notably, the conception of science featured there is a *dynamic* and *creative* one, opposed to the static conception which tends to emerge from science textbooks. Kuhn there takes what he variously calls 'preconceptions', 'prejudices', 'points of view', 'principles' or 'conceptual frameworks' to be essential to science, and the underlying notion here (perhaps closest to 'conceptual framework') is undoubtedly a precursor of the looser sense of his later term 'paradigm'. But the sense of 'paradigm' as *achievement*, which was to be absolutely central to SSR, wasn't yet present. The idea, which Kuhn derived from the work of the Swiss psychologist Jean Piaget (p. vi, RSS, pp. 279, 283), that although everyone operates within a certain 'perceptual' or 'behavioral world', different scientists operate within different behavioural worlds generated by their profession presages important ideas from SSR, including the concept of 'normal science'. So too does the idea that scientific activity usually consists in increasing the scope and precision of an existing system. The idea that such behavioural worlds confront 'anomalies' which sometimes change those worlds by worsening into 'crises', is also present, as is the idea of scientific revolutions, destructive phases in which existing systems are replaced by new ones.[3]

While still at Harvard, working first as an instructor and then as an assistant professor, Kuhn developed an undergraduate course on 'The Development of Mechanics from Aristotle to Newton'. Although he had set it up, Conant soon decided to stop lecturing on the history of science course. Along with the chemist Leonard Nash, Kuhn took over its teaching. The lectures he gave, on the transition from the geocentric world-view of Ptolemy to Copernicus' heliocentrism resulted in his first book *The Copernican Revolution* (1957). Though well received by historians of science, it caused none of the commotion that his later work was to generate, even though one of its main points was that 'the' Copernican revolution was *plural* (CR, pp. vii–viii).

Towards the end of his time at Harvard, Kuhn was invited by Charles Morris, one of the associate editors of the *International Encyclopedia of Unified Science*, to write its planned volume on the history of science, the originally envisaged author having dropped out. Kuhn agreed, and henceforth corresponded with the other associate editor, Rudolf Carnap, who enthusiastically endorsed the book's publication in the series, and served as its editor.[4] The *Encyclopedia* had been conceived in the 1930s by its editor-in-chief, Otto Neurath, as a collection of monographs which would illustrate the 'logical positivist' view of sciences. By the time SSR was commissioned, in the early 1950s, logical positivism had dissipated, since its advocates, mainly based in continental Europe, had been forced into exile by the rise of Hitler. Some of them then regrouped in the US to form the logical empiricist movement, and it was the leading figures of this movement who authored many of the *Encyclopedia*'s nineteen monographs.

Kuhn only really intended to use this publishing invitation 'to produce the first version, a short version of *The Structure of Scientific Revolutions*' (RSS, p. 292), but even this took far longer than he anticipated. He specifically says that SSR is 'an essay', an interim progress report, rather than 'the full-scale book my subject will ultimately demand' (p. viii). But that full-scale book was never to appear.

In 1956, Kuhn left Harvard for the University of California at Berkeley. Berkeley initially offered him a position in their Department of Philosophy, which appealed to him because he wanted to return to his philosophical interests. They then suggested that he hold a joint position in Philosophy and in History, and he agreed. Kuhn's distinctive idea of 'normal science' first appeared in his important formative article 'The Function of Measurement in Modern Physical Science', which was drafted in this same year, revised and extended in 1958, and eventually published in 1961.

After the publication of CR, Kuhn did indeed return to philosophy. During his time at Berkeley (1956–64), where he taught courses in the history of science, and in philosophy, he generalized the scheme he had already used to characterize the Copernican revolution. Among his colleagues in philosophy there, Kuhn befriended and had extensive discussions with Stanley Cavell and, from 1959, Paul Feyerabend. Cavell, in particular, helped Kuhn with reading Ludwig Wittgenstein's *Philosophical Investigations* (1953), a book which was encountering a rough reception in a department still dominated by the ghost of logical positivism, but with which Kuhn felt sympathy, as well as a certain impatience.

Kuhn spent time at Stanford University's Center for Advanced Studies in the Behavioral Sciences during 1958–9. He used this time for preparing SSR, and this seems to be when he forged his own concept of a paradigm, which first saw print in his 1959 lecture 'The Essential Tension'.[5] Kuhn took this concept to be 'the missing element I required in order to write the book' (ET, p. xix). This key concept, for him, denotes the specific *kind* of consensus characteristic of mature scientific disciplines.

Although actually writing SSR while at Stanford proved slow and difficult, Kuhn managed to do the same job quickly when he returned to Berkeley. He sent a draft version to Conant and to Feyerabend early in 1961. Conant's response manifested a worry that Kuhn was using the word 'paradigm' as a panacea. Feyerabend, who sent him several letters about the book (see Hoyningen-Huene 1995 and 2006), had even deeper concerns. But SSR, the first edition of which was dedicated to Conant, appeared in late 1962, as the last and longest of the *Encyclopedia*'s monograph series, as well as in the form of a separate book. This was the work that brought Kuhn to a worldwide audience. Having now sold more than a million copies, and having been translated into 27 different languages, it has become, as Richard Rorty put it, 'the most widely read, and most influential, work of philosophy written in English since the Second World War' (Rorty 2000, p. 204). From the start, it was received well by social scientists and by students intoxicated with the very idea of revolutions, but it also gradually began to attract far more critical attention from philosophers. Kuhn was never entirely happy with any part of this reception. His fans among the student revolutionaries, and many of those in the social sciences, misunderstood him. As he later put it, 'I was trying to explain how it could be that the most rigid of all disciplines, and in certain circumstances the most authoritarian, could also be the most creative of novelty' (RSS, p. 308). And although philosophers were the audience he most courted, he often considered them to have misread his work. Where their accusations were of irrationalism this produced annoyance or anger; where they were of relativism and/or idealism, his responses varied from pointed qualification to outright rejection. He tried further to explain his views in ways that would not incur these philosophical objections in a 'Postscript' to SSR, and then in important conference papers.

In the early 1960s, soon after finishing the manuscript of SSR, Kuhn was invited by a joint committee of the American Physical Society and the American Philosophical Society to construct an archive of information documenting the history of quantum physics, and did so in collaboration with some of his graduate students, John Heilbron, Paul Forman and Lini Allen. The project, which began in Berkeley, moved to Copenhagen, and then returned to Berkeley, ultimately issued in the book *Sources for the History of Quantum Physics* (Philadelphia: American Philosophical Society, 1967).

In 1964 Kuhn left Berkeley, whose philosophers seem to have appreciated him insufficiently, for Princeton, becoming the director of their History and Philosophy of Science programme in 1967, a full Professor of History in 1968 and a member of their Institute for Advanced Study in 1972. A first collection of his essays, *The Essential Tension: Selected Studies in Scientific Tradition and Change*, was published in 1977. But the main fruit of his work at Princeton was *Black-Body Theory and the Quantum Discontinuity* (1978), a formidable book on the early history of quantum mechanics, which surprised and disappointed many of its readers by not using the terms ('paradigm', 'normal science', 'scientific revolution', etc.) SSR had made famous. Kuhn, though, regarded it as the best and most representative of his historical works.

Kuhn left Princeton in 1978, moving first to New York University, and then joining the Department of Linguistics and Philosophy at the Massachusetts Institute of Technology (MIT), later becoming their Laurance S. Rockefeller Professor of Philosophy. Here his attention turned once again to philosophical matters, especially his concepts of scientific revolution and incommensurability, now more explicitly conceived as centring around language and meaning. He published more than half a dozen important articles during this time, extending, reviewing, refining or altering the ideas familiar from SSR.

Kuhn retired in 1991, becoming a Professor Emeritus, but published very little thereafter. He died in 1996, having been ill with cancer for two years. A second collection of his essays, *The Road*

Since Structure: *Philosophical Essays, 1970–1993*, was published in 2000. But the final phase of his work is still to come to fruition. His last book, *The Plurality of Worlds: An Evolutionary Theory of Scientific Discovery*, on which he worked for the last 15 years of his life, and which was two-thirds completed at the time of his death, is currently being edited for publication by John Haugeland and James Conant, the grandson of James Bryant Conant.

CHAPTER 2

OVERVIEW OF THEMES

Kuhn's path, then, took him from physics, to the history of science, and then on to philosophy (p. v). But although his training as a physicist was formal and complete to doctoral level, he was largely self-taught in both the history of science and philosophy. His principal *interest* was in the latter, but he wasn't always careful enough, as a philosophical thinker, to present a clear overview of what he was saying, or to avoid certain stock philosophical objections. There were some significant changes to his views, too, even in the time between the appearance of SSR's first edition (1962) and its second (1970).

Kuhn was reacting against the prevalent image of science, and meant SSR to be a sketch of a new image, a 'historically oriented view of science' (p. x) which was gradually emerging from the continental European historians of science. But SSR is somewhat in tension with itself as the result of being a collision between this continental 'historicist' approach to the history of science, informed by what Kuhn thought of as an emerging complex of results from psychology, linguistics and related disciplines, on the one hand, and a framework derived from Anglo-American 'analytical' philosophy, on the other. The resulting tensions are, as Ian Hacking has noted, what makes Kuhn's work great (Hacking 1979, p. 236). But they are also what ultimately threaten to tear it apart and mean we have to go beyond it. One way of doing so, which I will pursue here, is to try to see what can be done to excavate from SSR a coherent philosophical picture. For the above reasons, of course, any attribution of such a picture to Kuhn must be something of a reconstruction.[1] But that reconstruction is still worthwhile.

Fundamentally, Kuhn characterizes science in terms of its *mode of instruction* and its *development*. The instruction one receives when entering a scientific profession means that one learns to be a scientist and to practise a particular science in a particular *way*. Kuhn's study of the history of science convinced him that in the past of each science there has been more than one such way. This conviction that sciences *change*, even at this most basic level, means that Kuhn's picture of science is, famously, a picture of how sciences *develop*.

The practices of science, however, are constitutively connected with other phenomena. Although science involves *claims* (such as theories), these must be expressed in terms of something that is not a claim: something like a *conceptual scheme*. This is where Kuhn's concept of a *paradigm* comes in. Paradigms, in the sense which has become more popular, are *amalgams* of scientific practices and scientific conceptual schemes. They are ways of doing the science in question, but also ways of thinking. Kuhn was right to insist that neither practices nor conceptual schemes can be assessed in terms of truth-or-falsity. Their assessment must be in terms of different features.

Thus, Kuhn argues in SSR's section II (and as we shall see in this guide's section 2), each natural science begins with a pre-paradigm period, during which there is no agreed way of undertaking that discipline. Research done during this phase involves scientists, but it doesn't quite constitute *science*.

At some point, however, one of these researchers makes a breakthrough in explaining some handful of the heap of relevant phenomena. That person, in Kuhn's terms, invents a *paradigm*. For as long as that paradigm holds sway over that field, its scientists conduct what Kuhn calls 'normal science'. Normal science, which constitutes *most* scientific activity, is the first feature of science that Kuhn thought had been neglected, and to which he wanted to draw attention. (Paradigms and normal science are the subjects of SSR's sections III–V, and this guide's section 3).

In the course of their normal scientific research, though, scientists turn up certain problems, certain failures of fit between paradigm and experimental or observational results. These are what Kuhn (SSR section VI) calls 'anomalies'. If they persist, they can turn into 'crises' (SSR section VII), episodes in which the reigning paradigm begins to lose both its focus and its grip. But even when they do so, scientists don't react to crises in the sort of way that philosophers of science usually suggest (SSR, section VIII). Anomalies, crises, and how scientists react to them are covered in this guide's section 4.

A paradigm's grip is finally lost only when it is succeeded by a rival paradigm. That transition is the one Kuhn made famous as a *scientific revolution*. Such revolutions are the second feature of science to which Kuhn wanted to draw attention. Since the sheer prevalence of such episodes had seldom been suspected before, and Kuhn took various kinds of historians and philosophers of science to be committed to denying their existence or significance, SSR spends considerable time trying to convince its readers that scientific revolutions do exist (section IX), and explaining why they have been thought not to (section XI). This guide confronts Kuhn's discontinuous picture of science with the more familiar, gradualist picture known as *cumulativism* (section 5).

The existence of scientific revolutions is also supposed to have dramatic consequences for our general picture of science (SSR sections X, XII). The *relationship* between the old paradigm and its successor in any given field can be particularly problematic, amounting, Kuhn argues, to an '*incommensurability*' which upsets previous philosophical accounts of science. One aspect or consequence of that incommensurability is supposedly that in such a revolution, the world that scientists work in changes, in a way analogous to a 'Gestalt-switch'. Kuhn never doubted that science nevertheless makes *progress* (SSR section XIII), but his account of what this consists in had to be different from that of positivists, logical empiricists, critical rationalists and most 'scientific realists'. While he always thought that the mature natural sciences constitute our best-developed forms of knowledge, he never subscribed to the usual idea that they develop by coming ever closer to the truth, and he did his best to challenge *scientism*, the belief that science is a general intellectual panacea. Sections 6, 7 and 8 of this guide try to disentangle these issues in a way that suggests which aspects of Kuhn's views deserve development, and which merit rejection. In section 6, I try to identify an overall perspective that avoids certain philosophical problems that Kuhn recognized, and which I (following certain others) call 'conceptual relativism'. (That expression has been used before in characterizing Kuhn's views, but I mean something rather different from certain other Kuhn commentators.) While I don't take this to be a perspective Kuhn ever succeeded in *formulating*, I do think it makes sense of much of what he said, and was tempted to say, and that it merits development.

Kuhn added a very significant 'Postscript' to SSR for its second edition, in which he clarified some issues, reaffirming many of his views and altering a few others. The issues raised there are treated, in this guide, at the end of the sections to which they are most pertinent.

CHAPTER 3

READING *THE STRUCTURE OF SCIENTIFIC REVOLUTIONS*

SECTION 1. KUHN'S PROJECT AND ITS CENTRAL TERMS

Section 1 of SSR sets the stage by outlining the target of Kuhn's critical focus, his project, the context of his work and some aspects of the new image of science he hopes will result. Kuhn introduces some of the key terms in this new image, notably 'normal science' and 'scientific revolutions', but explains them only in a preliminary way.

'History, if viewed as a repository for more than anecdote or chronology, could provide a decisive transformation in the image of science by which we are now possessed' (p. 1). The ringing start of the first section of Kuhn's book announces both his project and his target.

The existing image of science Kuhn has in mind seems to have been derived from his own scientific training, from finished scientific achievements such as science textbooks, and from his acquaintance with certain writings on the history and philosophy of science. Although he takes little trouble to specify this target in SSR, he did later say a little more about where he got the image of science he was reacting against. During his war service he read books by, among others, the physicists and philosophers of science Percy Bridgman and Arthur Eddington. By 1945 he had read some of Bertrand Russell's philosophical works, and he later encountered the work of the logical positivists Philipp Frank, Richard von Mises and Rudolf Carnap.[1]

These works, which he 'took to be philosophy of science', together with a number of other 'quasi-popular, quasi-philosophical works' (RSS, p. 305) set his target. It was, as he put it, the *'everyday image of logical positivism*' he was reacting against (ibid., p. 306, emphasis added). His acquaintance with the philosophy of science was thus rather superficial and, as he later admitted, if he had known the philosophical literature in depth, SSR would probably never have been written.

Kuhn's target, then, even in SSR, was not any particular developed philosophical view of science (he seems to have been unacquainted, at that time, with the *developed* views of either of the two main contemporary approaches to the philosophy of science, logical empiricism and Popper's 'critical rationalism' (see RSS, p. 227)). Nevertheless, he quite plausibly considered this an image by which people were *generally* possessed, being 'part of the ideology of scientists' (RSS, p. 282), and the kind of image that filters through to the general public.

Kuhn's complaint was principally that this image of science simply didn't stand up to historical scrutiny. This is for several reasons. Scientific texts, Kuhn himself argues, are necessarily unhistorical, since what a student needs is to become skilled at using the *current* methods, not the old ones. The questions prompted by science textbooks are unhistorical, and make it impossible to understand the development of science. The sort of history that emerges from science textbooks is 'not quite history' (RSS, p. 282). Science never forgets its heroes, but it *does* forget how they, and the scientific communities of which they were members, came to achieve what they did. Instead of looking at science as a set of *products*, as most scientists and philosophers, and even many historians had done, Kuhn follows the newer historians and looks at science as a *process*, a process which includes all sorts of phenomena that never make it into the finished (published) products of science.

A second reason why the existing image didn't fit the history of science, though, was that the philosophers of science involved conceived their own discipline as a *normative* activity. The logical positivists, for example, at least in their better-known phase, weren't even *trying* to give a historically accurate image of science.[2] Their project was the rather different one of giving a 'rational reconstruction' of the *logic* of science, the relations between different kinds of scientific statements.

Popper and his critical rationalists (at that time including Kuhn's Berkeley colleague and friend Paul Feyerabend) tried harder to make contact with the history of science, and were more sympathetic to the works of the new historians of science Kuhn admired, but they still conceived of their project in strongly normative terms. For them, methodological rules are rules laid down for scientists by philosophers (see Chapter II of Popper 1959). They aimed to say what science is *at its best*, rather than giving an accurate picture of how science really and *typically* is. Indeed, they insisted that *any* account of the history of science will be informed by a selection of material which is based on certain values, and that *different* values would result in different (but still historically accurate) accounts of the history of science.

Kuhn had a different conception of the philosophy of science. He insisted that the evaluation of activities can only follow their accurate description. We would have to have an accurate *picture* of science before we could say whether any given activity counts as scientific or not, and before we could evaluate such activities. He therefore proposed, on broadly empiricist lines, to take a closer look at science, assuming that doing so would narrow down the range of views one might take of its history. He thought of himself as giving 'a more realistic appraisal of scientific theories'.[3] Kuhn's book was one of the first attempts to describe, in a general way, what most of those employed as scientists spend most of their professional time doing, and thus to make more sense than had been made before of more of the activity which we all pre-theoretically recognize as science. There is a worry, though, that making such sense amounts to '*legitimating*' science as a whole, the activities of *typical* scientists. Critical rationalists resist this. For them, it may or may not be that typical scientists proceed correctly. Whether they do so depends on whether or not they meet an external, universally applicable standard (derived from philosophy). Science may have taken a wrong turn, losing sight of its own nature.

SSR did improve on current views in presenting science as a set of skilled *activities* or *practises*, not just a system of *statements*. (The logical positivists had explicitly committed to such a view, the critical rationalists' focus on deduction led them in the same direction, and even Koyré's approach had a similar effect, being thoroughly 'intellectualist'.) Kuhn thereby also distanced himself from any *formalist* view of science.

He also presented an image which runs counter to the logical positivist/logical empiricist idea of *the unity of science*. The positivists undoubtedly conceived of this unity in different ways. Carnap, for example, initially conceived of it as unity by virtue of reduction, the concepts of 'higher-level' sciences being statable in terms of the concepts of 'lower-level' sciences such as physics. Otto Neurath, however, merely thought the sciences share a common ('physicalistic') *language*. Kuhn's image is opposed to all such conceptions, though, since he insists not only that there is a *succession* of different conceptual frameworks in any given scientific field, but also that these successive frameworks need not even address the same problems.

Kuhn's image also opposed the positivist and empiricist tendency to *separate* science from metaphysics. Kuhn followed Koyré's determinedly *philosophical* approach to the history of science in recognizing the constitutive role of metaphysics in the history of science. Following the idea that philosophic trends have an essential influence on scientific theories (Koyré 1954), one cannot help concluding that metaphysics is somehow inseparable from science. Popper, too, had already recognized this. But logical positivists, logical empiricists and inductivists of other kinds are united in opposing the idea. (Followers of Wittgenstein also have trouble with it.) The positivist idea that one might once and for all declare a proposition meaningless is incompatible with Kuhn's idea that scientific progress is punctuated by crises in which meanings are renegotiated.

Kuhn identified the main feature of the received image of science as *cumulativism*, the view that science advances by piecemeal accumulation. The resulting task of the historian of science is to chronicle the advances (discoveries) and obstacles to this cumulative progress. But the continental historians of science whose work he knew well found it more and more difficult to work in this way.

Kuhn, then, undoubtedly perceived himself as opposing the logical positivist/logical empiricist view, which he thought of as cumulativist, formalist and ahistorical (if not *anti*historical). But although this perception contains some truth, it must be heavily qualified.

For one thing, the earliest logical positivists (Neurath, Frank and Hans Hahn), influenced by the French conventionalists, *had* taken a substantial interest in the history of science, and were not committed to either formalism or cumulativism.[4] One is tempted not to blame Kuhn for not knowing much about this phase of positivism, which has been studied only recently. But Frank, some of whose work Kuhn says he knew, and who was at Harvard at the same time as Kuhn, was involved in this early phase, published an account of it in 1949, and also edited a volume (Frank 1954), one section of which is devoted to 'Science as a Social and Historical Phenomenon', and which includes an essay by Koyré which Kuhn would almost certainly have known.

For another thing, some central features of Kuhn's view had been anticipated, albeit in a different key, as it were, by Carnap himself. The image of Kuhn as radically opposed to logical empiricism needs to be tempered with the realization that Carnap had already highlighted some of the 'practical' aspects of scientific reasoning (see Reisch 1991, p. 276). Kuhn, who admitted to not knowing Carnap's more recent work at the time SSR was published, and who didn't take Carnap's friendly editorial encouragement as evidence of any serious consonance between their views, seems eventually to have been persuaded of this consonance (RSS, pp. 227, 305–6). But he continued to insist that for Carnap the importance of language change was 'merely pragmatic', and not *cognitive* (ibid., p. 227).

On the horizon, Kuhn saw the possibility of a *new* image or 'concept' (p. 1) of science, derived from the work of the historians he admired. These were the people, led by Koyré, then involved in the '*historiographic revolution*', a change in the way the history of science is written. These historians, Kuhn argued, found it increasingly difficult to answer the sorts of questions that the usual image of science prompts them to consider. (Later he will suggest that such questions are not merely intractably difficult, but that, having no answer, they're confused or *ill-formed*.) When investigating intellectual products very different from our contemporary scientific theories, instead of having initially clear demarcations of science from error and superstition *confirmed* as their work went into more and more depth, these same historians became more and more confident that the products they were investigating were in no way less *scientific* than today's theories.

These problems should make us reconsider the usual image's idea that science advances by the accumulation of piecemeal results. If, as the historian is forced to say, science does include 'bodies of belief quite incompatible with the ones we hold today' (p. 2), cumulativism must be wrong. It has to be said that this might well seem to be news *only* to those deriving their views of science from science textbooks or the popular press, not to any historian *or* philosopher of science!

Few twentieth-century *historians* of science thought of themselves as chroniclers of piecemeal accumulation. And most philosophers of science in the first half of the twentieth century were familiar with the idea that the change from the Newtonian world-view to that of Einstein's theory of relativity, a transition they themselves had lived through, constituted a revolution in physics. This confirms that Kuhn's target was a *very* naive cumulativism. As we shall see in section 5, though, cumulativist approaches to the history of science can and do still have important defenders.

The historiographic revolution prompted new sorts of questions about science, and resulted in a very different image of the *development* of science. This new image was driven by a historiography which, '[r]ather than seeking the permanent contributions of an older science to our present vantage, attempt[s] to display the historical integrity of that science in its own time' (p. 3). The historian must now try to present older views as being as *internally coherent* as the sources allow. This historically sensitive approach (related to the kind of 'interpretative' social science familiar from the post-Kantian philosophical tradition) would enable us to see that scientific methodology doesn't dictate a single answer to all important scientific questions; that observation and experience don't narrow the field down to a single body of belief; and that effective scientific research can't begin before the scientific community has decided on answers to questions about explanation, observation and ontology (pp. 4–5).

The idea that there is such as thing as 'the scientific community', seems to have been introduced by the Hungarian philosopher-scientist Michael Polanyi, as early as 1942.[5] But, as with the concept of scientific revolutions, Kuhn's innovation was to talk not just about 'the scientific community' (e.g. p. 4), but also about particular scientific *communities* (ibid., also pp. 11, 49, 177). As we shall see, the communities he has in mind can be small, 'consisting perhaps of fewer than twenty-five people' (p. 181). However, Kuhn wasn't always careful about which characteristics are those of individual scientists, and which are those of the scientific community (or of a *specific* scientific community) (ET, p. 227 note).

Kuhn's project, then, was to summarize, popularize and develop the historiographic revolution, and to indicate how it (together with apparently related developments) might transform both the *philosophy* of science and our general image or *concept* of science. It

19

was, as he put it, *'primarily a work of synthesis'*.[6] He didn't think of himself as working on his own, or as breaking entirely new ground, but as drawing out, making explicit, the implications of a new kind of history of science, and a new kind of philosophy of science, which were already under development (p. 3). While most of Kuhn's *historical* influences (Koyré, Meyerson, Metzger) pre-date even the conception of SSR, the works in the philosophy of science Kuhn was drawing from appeared during SSR's *gestation*, not before it was conceived.

Kuhn developed a *general* sketch partly because the work of the new historians hadn't yet lent itself to any such overview, and partly because only such a thing could displace the existing 'image' of science. SSR was intended as an outline of the new image of science, a salvo against a narrower conception present in the best-known phase of logical positivism which chimed with certain every-day assumptions about science, and an attempt to dislodge a *philosophical* picture initiated by the scientific revolution of the seventeenth century. In the Postscript to SSR, Kuhn took pains to point out that the book's developmental portrayal of science borrows its themes from the work of historians. Applying concepts such as structure and revolution, already familiar from the field of history, *to science*, is one of only two respects in which he very modestly deemed SSR original (p. 208).

Kuhn also, in SSR's first section, flags some of the conclusions he will come to. We shall note these as and when they occur in the rest of his text. But here it's worth singling out a question he raises about his own investigation, whether historical study can possibly effect the change he envisages, which he calls a *conceptual transformation* (p. 8). Dichotomies routinely deployed within the philosophy of science, such as those between the descriptive and the normative, the empirical and the logical, and the contexts of discovery and of justification, suggest not. But, these dichotomies, Kuhn feels, are themselves parts of the philosophical picture that will need to be *overcome* in the course of his investigations. As he suggests, the history of science couldn't fail to be 'a source of phenomena to which theories about knowledge may legitimately be asked to apply' (p. 9).

Kuhn's central idea is that the career of most sciences has a *typical pattern* which can be divided into historical phases or stages. Before he can describe the first such phase, he has to start by introducing three new and related concepts which characterize later phases of this pattern: *normal science, scientific revolutions* and *paradigms*.

Normal science is what almost all scientists spend almost all of their professional time doing. It is what is enshrined in the scientific textbooks of the time, for it is what must be taught to each new generation of scientific initiates. And it is based on what Kuhn calls 'paradigms' (the subject of our next section). But normal science is disrupted and transformed by occasional episodes which Kuhn calls *scientific revolutions*. The crucial thought here is that '[t]he successive transition from one paradigm to another via revolution is the usual developmental pattern of mature science' (p. 12).

Study questions

1. Did Kuhn choose too easy or broad a target for his critique? Was he attacking an image only the general public would (in the early 1960s) have endorsed? Or are there historians and philosophers of science who fit his characterization of the old image of science? How much does his characterization fit the views of the logical positivists, the logical empiricists, or Popper's 'critical rationalism'?
2. Was Kuhn being too naively empiricist in assuming that there can be such a thing as a more accurate description of actual scientific practice?

SECTION 2. THE PRE-PARADIGM PERIOD

Section II of SSR introduces another key term in Kuhn's image of science, 'paradigm'. Initial characterizations of paradigms, their creation, and their roles are given, as are some examples of paradigms. These are in the service of contrasting normal science, which consists of research under a paradigm, with the earlier, pre-paradigm period, whose nature and demise are then described.

Normal science consists, Kuhn says, in 'research firmly based upon one or more past scientific achievements, achievements that some particular scientific community acknowledges for a time as supplying the foundation for its further practice' (p. 10). The concrete achievements in question, which Kuhn calls 'paradigms', have to be unprecedented enough to attract an enduring group of followers, but at the same time open-ended enough to leave those followers the right kind of work to do.

Such achievements are what one studies when being educated to enter a scientific community. (They need not be *singular*: Benjamin Franklin's work, for example, provided those researching into the nature of electricity with a paradigm by explaining many of the effects other theorists could (separately) explain.) Researchers who share such a 'paradigm', Kuhn tells us, also share certain rules and standards for scientific practice. As a result, there's a kind of *consensus* about fundamentals which is necessary for normal science. But what *underlie* (and are prior to) the other aspects of paradigms (concepts, laws, theories, etc.) are the *achievements* in question.

Before Kuhn got hold of the term 'paradigm' and, by his own admission, ruined it (RSS, p. 298), it had an established ordinary meaning, as well as both a technical and at least one philosophical use. Its ordinary meaning, deriving ultimately from Greek, is a pattern or example. In the teaching of grammar, it was a technical term used to mean a standard and basic example of a word's inflexion (the conjugation of a verb, or declension of a noun). In philosophy of science, the term has a long history that goes back to the eighteenth-century German scientist and philosopher Georg Christoph Lichtenberg.[7] He applied the term to science and scientific change, but (unlike Kuhn) explicitly took on board its technical grammatical use, applying it to scientific achievements which function analogously to grammatical standards, as models for the solution of problems which obviate the need for explicit rules. Wittgenstein, who was influenced by Lichtenberg's work, used the German term '*Paradigma*' in his lectures for something like a model or stereotype. He also used it in his *Philosophical Investigations*, which Kuhn had read in 1959, for 'something with which comparison is made' (Wittgenstein 1958, §50, see also §§51, 55, 57, 215, 300), but it's not clear that Wittgenstein's usage departs from the ordinary use in the direction of Lichtenberg's semi-technical one.[8] Logical positivists such as Neurath and Moritz Schlick, as well as philosophers of science influenced by Wittgenstein, such as W. H. Watson, Norwood Russell Hanson, and Stephen Toulmin, were using the term from the 1930s onwards, albeit in a way indistinguishable from its ordinary meaning. Kuhn himself used the term in this informal way in CR, too.

In philosophy more generally, again under the influence of Wittgenstein and the Oxford linguistic philosophers, some were persuaded that the use of standard examples from which one catches on to the meaning of a word licensed what was known as 'paradigm-case arguments'. In such arguments, there being a paradigm for our use of a word is supposed to establish the existence of whatever the word supposedly refers to. Such arguments are no longer in good odour, and the concept of a paradigm may not solve philosophical problems. Nevertheless, the idea of *a standard example from which one catches on*, to pick out features of science, need not be impugned.

However, critics were quick to notice that Kuhn used the term 'paradigm' very widely and in rather different ways. Kuhn himself later conceded that within SSR he had used the concept in two main senses, which he distinguished in his Postscript (and in ET, p. xix–xx). It is important to distinguish the different things Kuhn means by 'paradigm', since some of the things he says about paradigms make no sense when applied to the wrong kind of paradigm. (Likewise, as we shall see, it's important to distinguish small from large scientific revolutions.)

The first sense of 'paradigm' is the one we have already met, in which it means a *concrete achievement* or *model* from which initiates are taught. For this sense of 'paradigm' Kuhn came to use the term *'exemplar'* (short for 'exemplary achievement'). He took this to be the more fundamental sense, and developing it was the one other respect in which he deemed SSR original (p. 208).

But it's notable that a second sense of the term (if not something even *more* fuzzy) is the only one that really came to public consciousness. This is what Kuhn later called a *'disciplinary matrix'*, meaning a larger, more encompassing cognitive structure, 'the entire constellation of beliefs, values, techniques and so on shared by the members of a given [scientific] community' (p. 175).[9] Kuhn later took himself to task for using the term 'paradigm' in this way, for 'a hell of a lot of other things that weren't models'. This, he said, 'made it very easy to miss what I thought of as my point entirely, and to simply make it the whole bloody tradition, which is the main way it has been used since' (RSS, p. 299).

In introducing the concept of a paradigm, Kuhn said he wanted to suggest that 'some accepted examples of actual scientific practice . . . provide *models* from which spring particular coherent traditions of scientific research' (p. 10, emphasis added). However, it isn't always possible to tell from the context *which* kind of example of scientific practice, and thus which sense of 'paradigm', Kuhn had in mind. (It's *already* somewhat ambiguous, for example, within the first two pages of its introduction, as between an achievement *itself*, and the complex of this achievement *plus* laws, theories, etc.)

Some commentators, having noted this, persist in using 'paradigm' indiscriminately (sometimes on the ground that, having given birth to the term's new use, Kuhn was no longer in control of it). But this can be confusing, and prevents detailed evaluation of SSR's claims. Instead, where one *can* work out that Kuhn had one rather than another sense of the term in mind, one should try to follow his later suggestion of distinguishing them.

Some take Kuhn to be using the terms 'paradigm' and 'scientific revolution' only for large-scale scientific phenomena. This is wrong, since, not just within but also before SSR, he clearly applies both expressions to small aspects of science, as well as to large ones. (He explicitly discusses small scientific revolutions such as the discovery of X-rays, as well as large ones such as the Copernican revolution.)

In fact, most of the examples of paradigms from SSR's first edition fall into the category of *exemplars*.[10] He talks, for example, of 'the Franklinian paradigm' (p. 18), by which he means Franklin's successful *explanation* of the Leyden jar, which 'made his theory a paradigm' (p. 17). (Both a theory and an explanation can be an achievement.) A synthesis successful in attracting the next generation of practitioners is also said to count as a paradigm (p. 18), the kind of thing that transforms pre-paradigm research into a science. And later on, Aristotle's analysis of motion, Ptolemy's computations of planetary position, Lavoisier's application of the balance, and Maxwell's mathematization of the electromagnetic field are all achievements explicitly said to be paradigms (p. 23). Paradigms are said to be *qualitative* rather than quantitative (p. 29).

However, Kuhn refers also to 'the corpuscular paradigm' (p. 105), by which he seems to mean the 'mechanico-corpuscular world view'. This would be 'paradigm' in its 'disciplinary matrix' sense.

An important aspect of Kuhn's scheme that isn't always picked up on, is that commitment to a given paradigm represents *a particular way of doing science*. Kuhn thus recognizes that scientists' commitment always includes 'an important element of the historically accidental, the temporarily local, and, thus, of the arbitrary' (Kuhn in Crombie 1963, p. 393). However impressive any particular paradigm seems, there are always *other* possible ways of doing science.

Two philosophers who certainly have picked up on this are Stanley Cavell and Ian Hacking. Cavell recognizes Kuhn's focus on the '*tacit*', or '*subliminal*' aspects of science, and he connects scientific revolutions with changes in what Wittgenstein called '*natural reactions*': 'Perhaps the idea of a new historical period is an idea of a generation whose natural reactions – not merely whose ideas or mores – diverge from the old; it is an idea of a new (human) nature' (Cavell 1969, p. 121). It is tempting to contrast this idea of different groups of scientists having different natures with Popper's idea that we humans are creatures that let our theories die in our stead. Cavell doesn't deny that one scientist can have *more* than one set of 'natural reactions', but he might insist that this is the exception, rather than the rule, when it comes to scientific *paradigms*.

Ian Hacking, for his part, has pursued Kuhn's ideas by focusing on the idea that different groups of scientists deploy different *styles* of scientific thinking, styles which themselves determine the sense of scientific propositions (see, for example, Hacking 1985).

The first phase in the typical pattern of the history of a science, which forms most of the subject matter of Kuhn's section II, is the *pre-paradigm* or *pre-consensus* period. Such periods are marked by a certain kind of thorough-going *diversity* or *fragmentation*. There are multiple views of the phenomena in question, amounting to different *concepts* of those phenomena. There are no agreed canons of explanation, no fixed methodology, no acknowledged scientific authorities. Instead there's a plurality of competing 'schools' of doctrine, each of them deriving strength 'from its relation to some particular metaphysic' (p. 12), but none having the upper hand. The activity of the researchers involved allows for unlimited disagreement and the criticism of each and every assumption. But this debate over fundamentals is directed against other researchers, not towards nature. Even when there is a single theory of the domain, there's no single *interpretation* of that theory, no agreement on its achievements, methods, problems, or hopeful lines of solution. Instead there is what Kuhn comes to call a *proliferation of versions* of the theory. Observations which conflict with the theory are dealt with in an *ad hoc* way, or are simply ignored.

The consensus that paradigms involve is attained only with great difficulty, Kuhn suggests. This is because before a field has a paradigm all the facts its study might involve may well seem equally relevant. Research in these periods approximates most closely the kind of random 'fact-gathering' that naive empiricists like Francis Bacon are supposed to have held to constitute science. Although this can be essential to the *origins* of a science, it produces not a systematic body of knowledge, but what Kuhn calls a *morass*. The data collected are usually those that lie closest to hand, on the surface, and they are often collected by amateur observers, since there's no motivation for experiments in which nature is asked *detailed* questions. Much effort is expended over natural histories which 'classify' specimens and data, where in a later period these kinds of classifications would be pre-determined by the paradigm itself. In the absence of a paradigm, researchers have no system to tell them what details are relevant and irrelevant, thus the facts collected remain in a heap, as it were, rather than a structure. Each writer has to reconstruct the field from its foundations, since no body of belief can be taken for granted. Such writers describe and interpret the same facts, but their descriptions and interpretations are different.

Kuhn seems somewhat unclear on whether science before paradigms is possible, and on whether each science, properly so-called, is constituted upon a *single* paradigm. On the one hand he says that there can be a *kind* of scientific research without paradigms (exemplars), and that exemplars are a sign of the *maturity* of a scientific field (p. 11). His examples of pre-consensus periods are: the study of optics before Newton, the study of motion before Aristotle, the study of statics before Archimedes, the study of heat before Joseph Black, chemistry before Robert Boyle and Herman Boerhaave, and the study of geology before James Hutton, and he grants that such research involves *scientists*.

On the other hand, Kuhn observes that although the greatest figures in these periods *are* scientists, the net result of their activity is 'something less than science' (p. 13), and he counts the period before Newton as a pre-consensus phase of optics because there was no *single* generally accepted view of light at the time.[11] This is to take the paradigm as what *constitutes* a science. (Kuhn later changed his mind about this, as we shall see. But in any case his vacillation may simply reflect the fact that the concept of science and its cognates don't have *sharp* boundaries.)

The pre-consensus period usually ends in a fairly sudden and decisive way, when one school becomes dominant. Given the disarray Kuhn thinks characterizes such periods, of course, one might well wonder *how* such dominance is attained. Kuhn doesn't say much about this, but it's clear he thinks one pre-consensus school triumphs by emphasizing only *some* of the morass of collected facts.

Here Kuhn begins to *develop* the concept of a paradigm in a subtle but unacknowledged way. His mentioning that a theory can be accepted as a paradigm (p. 17), together with his later talk (pp. 26–8, 53, 61) of 'the paradigm theory' introduce the possibility that a theory might (sometimes) *be* a paradigm. Although he doesn't explain this, presumably it just means that the exemplary achievement in question can be (although it need not be) a theory. The exemplar is a *successful* theory, a theory which constitutes an *achievement* because it successfully explains some range of phenomena. When such a theory seems markedly better than its competitors, by virtue of being generally accepted as having explained more phenomena than they do, it *becomes* a paradigm or comes to be used *as* a paradigm. (This doesn't mean that one can *generally* identify paradigms with theories, though. As we shall see, there are good reasons to resist that identification.)

The onset of a paradigm has several effects, although all of them fall under the rough heading of *specialization*. Intellectually, it gives researchers in the field confidence to *direct* their investigations; this increases the effectiveness of their research. Socially, it affects the structure of the research field, partly by converting new members to the cause of the exemplar, and partly by ensuring that those unconverted are no longer considered as belonging to the profession, their work being ignored. It can even help to transform a field *into* a profession, bringing with it learned societies, journals and a place in the teaching curriculum. As a result, although the *research* is now focused on nature, rather than on other schools, the *communications* of these researchers come to be directed more and more to each other, rather than to a public audience, or to the unconverted. The textbooks and journals marking the discipline are directed to its own professionals. Finally, the field in question comes to be conceived more narrowly and rigidly by its researchers. This means they can take its paradigm for granted, no longer feeling the need to build the field anew from its foundations.

This specialization, and the increasing autonomy which separates professionals in a given scientific field from their colleagues, as well as from the public, are often deplored, of course.[12] Kuhn, however, suggests they are essential preconditions of the *progress* of science. (We shall consider his suggestion later, when we come to his discussion of scientific progress.)

Kuhn closes this section by taking the onset of a paradigm as the sign that a field has become a science: 'Except with the advantage of hindsight, it is hard to find another criterion that so clearly proclaims a field a science' (p. 22). We make more sense of this, I think, if we take it as saying that an *exemplary achievement* (rather than a disciplinary matrix) is what constitutes a science. Kuhn considers the *kind* of achievement in question to be pretty much unique to science (whereas disciplinary matrices certainly seem to be the kind of thing one also finds in other fields). But his qualification about the advantage of hindsight is important, too. It means Kuhn thinks of himself only as trying to say how one might recognize a science *at the time in question*, rather than as trying to give a once-and-for-all criterion that would 'demarcate' science from pre-science (let alone from error, superstition, metaphysics, non-science, pseudo-science or nonsense, as some philosophers have tried to do).

In his Preface Kuhn conceded that his distinction between pre-consensus and post-paradigm periods is 'much too schematic' (p. ix). In fact, in his 1969 Postscript he admitted to wanting to revise SSR in this respect. The transition from the pre-science period to normal science, he concedes there, 'need not be associated with the first acquisition of a paradigm' (p. 179). Pre-scientific researchers already share paradigms, but the nature of those paradigms is such that normal-scientific research isn't yet possible. The paradigms in question aren't yet mature enough to provide the sort of guidance that paradigms give to normal scientists. Despite this concession, though, Kuhn *always* resisted blurring the distinction between normal and revolutionary periods.

Study questions

1. Why should we think there *is* a typical pattern in the history of a science? Can such a pattern form part of our *concept* of science, or would it be merely a *contingent* feature of science?
2. Which aspects of science count as parts of 'paradigms', and which don't? What are the possible relations between theories and paradigms? Is the existence of a paradigm a criterion of a field's scientific status?
3. Is science before paradigms possible? Why should detailed scientific investigation prove possible only under a paradigm?

SECTION 3: PARADIGMS AND NORMAL SCIENCE

Sections III–V of SSR characterize paradigms and normal science further. Kuhn asks: what do scientists do when engaged in 'normal science', and why? Paradigms turn out to be largely promissory notes, and normal science their redemption. Kuhn then characterizes the problems that normal science consists in. Section III describes three 'foci' for the investigation of problems surrounding normal scientific facts, and three kinds of theoretical problems of normal science. Section IV goes on to characterize normal science in terms of the rules which govern what Kuhn calls its puzzle-solving activity. Section V investigates the relations between paradigms, rules and normal science. Kuhn argues that the rules philosophers of science usually focus on, which should be fully articulable, are in fact secondary to something that, although not fully articulable, can more easily be identified: a ground consisting of paradigms.

The second phase in the historical development of a science is embodied in what we have already seen Kuhn calls, perhaps somewhat provocatively, *normal science*. If a paradigm really is an *achievement* (an exemplar), and thus something that has been done 'once and for all' (p. 23), what is there left for scientists to do when engaged, as they almost always are, in 'normal science'?

At this point, Kuhn enters a bit further into the explication of his concept of a paradigm. One aspect of the established use of the term (according to which paradigms are accepted models or patterns used by being repeated in various ways within a training process) he regards as misleading. A paradigm (exemplar) *in science*, on Kuhn's usage, 'is an object for further articulation and specification under new or more stringent conditions' (ibid.). This is because, when it first appears, the paradigm is *restricted* in both scope and precision. It gained its adherents because of its advantage in solving certain generally recognized and acute problems more successfully than its competitors. But this *doesn't* mean that it is anywhere near completely successful. The 'success' of a paradigm, says Kuhn,

is at the start largely a *promise* of success discoverable in selected and still incomplete examples. Normal science consists in the *actualization* of that promise, an actualization achieved by extending the knowledge of those facts that the paradigm displays as particularly revealing, by increasing the extent of the match between those facts and the paradigm's predictions, and by further articulation of the paradigm itself. (pp. 23–4, emphases added)

This means that most scientists, throughout their careers, are engaged in what Kuhn rather disparagingly calls 'mop-up work', attempts to fit nature into the 'preformed and relatively inflexible box that the paradigm supplies' (p. 24). What he will soon call '*anomalies*', phenomena that don't fit the paradigm's conceptual categories, aren't sought for, and often aren't noticed. Neither the discovery of new sets of phenomena nor the invention of new theories are parts of the aims of normal science, since that has a 'drastically restricted vision' (ibid.). But this restriction, which might well appear to be a defect, is *essential* to scientific progress, since it forces scientists to investigate a limited range of phenomena in a depth and detail that would be impossible without it. Like the establishment of a paradigm in the first place, part of this kind of problem-solving work represents a *permanent* achievement.

This is one respect in which Kuhn's image of science is simply intended to give a more realistic *description* of the professional activities of those people employed as scientists than the images of scientific activity which emerge from philosophy. His comment to the effect that few who aren't practitioners of a mature science realize how fascinating such 'mop-up work' can be might well be directed to philosophers of science, many of whom would construe such work as mere drudgery.

Normal science, according to Kuhn, has two principal components: experimental and observational 'fact-gathering', and theoretical activity. The former is focused upon three kinds of problems. First are those facts which the paradigm has shown to be 'particularly revealing of the nature of things' (p. 25). Since the paradigm has already employed them in solving problems, they are worth determining 'both with more precision and in a larger variety of situations' (ibid.). Much normal science therefore consists of attempts to increase the accuracy and scope with which such facts (e.g. positions, magnitudes and movements of astronomical objects, properties of elements, materials and phenomena) are known. Usually, complex special apparatus is needed for the task. Plenty of famous scientists owe their reputations to work of this kind, work which Kuhn calls the 'redetermination of a previously known sort of fact' (p. 26).

Second, there are those facts which can be compared directly with the predictions of what Kuhn calls 'the paradigm theory'. Even in the case of our best theories, there are usually few such facts. And even where there are such facts, theoretical and instrumental approximations often limit the agreement which can be expected. But '[i]mproving that agreement or finding new areas in which agreement can be demonstrated at all presents a constant challenge to the skill and imagination of the experimentalist and observer' (p. 26). Again, complex and special equipment is usually needed, and reputations can be made. This kind of work is even more obviously paradigm-dependent than the first kind, since the problem to be solved is set only by the paradigm, and 'often the paradigm theory is implicated directly in the design of apparatus able to solve the problem' (p. 27, see also Kuhn in Crombie 1963, p. 389).

The third and final kind of fact-gathering activity involved in normal science, but the most important, is the empirical articulation of the paradigm theory, which resolves its ambiguities and permits the solution of 'problems to which it had previously only drawn attention' (p. 27). Depending on what science is in question, such work may involve the precise determination of physical constants; or of quantitative laws; or of new ways of applying the paradigm. Of the three kinds of fact-gathering activity, this is the least clearly empirical.

The *theoretical* work involved in normal science can also be subdivided into much the same classes. Part of it consists of 'the use of existing theory to predict factual information of intrinsic value' (p. 30). But, Kuhn thinks, scientists generally regard this as 'hack work to be relegated to engineers or technicians' (ibid.).

A second class, which does appear in scientific journals, results from the difficulty of developing points of contact between the paradigm theory and nature. It comprises 'manipulations of theory undertaken, not because the predictions in which they result are intrinsically valuable, but because they can be confronted directly with experiment' (ibid.). Such manipulations are intended to 'display a new application of the paradigm or to increase the precision of an application that has already been made' (ibid.). In the more mathematical sciences, says Kuhn, most theoretical work is of this sort. Kuhn's attention to this aspect of the complexity of science, which one wouldn't have suspected from logical empiricism or Popper (or from contemporary 'scientific realism') is an important reminder on his part. Scientists *do* have to spend time clarifying and reformulating their theories, since it's not always obvious how they *apply* to nature. Points of contact between a theory and nature can be very difficult for scientists to, as Kuhn puts it, 'develop' (p. 30).

Finally, the third kind of theoretical work implicated in normal science involves paradigm articulation. This includes attempts to clarify a paradigm by reformulating or reinterpreting it. It dominates during periods when scientific development is predominantly qualitative.

The overwhelming majority of the research problems undertaken by even the very best scientists, according to Kuhn, usually fall into one of these three categories. Such problems, as we have just seen, almost never aim at producing novelties of fact or theory. Normal science does add to the scope and precision with which the paradigm can be applied. But why are scientists so devoted and enthusiastic in tackling these research problems?

Kuhn's answer is that normal science consists in *puzzle-solving*. Normal scientists often know what the outcomes of their experiments, calculations, etc. should be, but they still devote their energies to showing *how* those outcomes can be achieved:

> Bringing a normal research problem to a conclusion is achieving the anticipated in a new way, and it requires the solution of all sorts of complex instrumental, conceptual, and mathematical puzzles. (p. 36)

A puzzle, Kuhn tells us, is a particular *kind* of problem that tests ingenuity or skill in its solution. Where problems may have no solutions, though, puzzles must have, although these need not be intrinsically interesting or important, and there must be 'rules' which limit the nature of acceptable solutions and the ways in which they can be obtained. Paradigms provide criteria for choosing just such puzzles. These are, Kuhn says, to a great extent, 'the only problems that the community will admit as scientific or encourage its members to undertake' (p. 37). One powerful way in which paradigms restrict vision is by ensuring that other problems are rejected as metaphysical, or as the concern of some other discipline, or as a waste of time:

> A paradigm can . . . even insulate the community from those socially important problems that are not reducible to the puzzle form [e.g. finding a cure for cancer (p. 36)], because they cannot be stated in terms of the conceptual and instrumental tools the

paradigm supplies. [. . .] One of the reasons why normal science seems to progress so rapidly is that its practitioners concentrate on problems that only their own lack of ingenuity should keep them from solving. (p. 37)

This is supposed to explain the passion with which scientists attack puzzles during normal science. Once a scientist, one is challenged and spurred on mainly by the thought that one will succeed in solving a puzzle which no one has solved before, or solved as well. Those scientists who do succeed in this prove themselves to be expert puzzle-solvers.

Here occurs one of the strongest contrasts between Kuhn's image of science and those associated with Popper and Feyerabend, who both portray scientists as far more than puzzle-solvers. Popper, for example, started from the thought that we can get a simplified picture of science from looking at the achievements of great scientists such as Galileo, Kepler and Newton. This starting-point obviously represents something of an idealization, for in focusing on the work of great scientists, one might well be ignoring the work of *most* scientists. Popper admitted as much, explaining that he wanted to convey 'a heroic and romantic idea of science and its workers: men who humbly devoted themselves to the search for truth, to the growth of our knowledge; men whose life consisted in an adventure of bold ideas' (Popper 1974, p. 977). Feyerabend, too, tended to discuss individual scientific heroes. Kuhn, though, was trying to represent *most* scientific activity, not merely the occasional, heroic and potentially revolutionary aspects of science. His concern was with science as a *profession*, rather than science as a purely intellectual adventure. Popper and Feyerabend, by contrast, simply weren't interested in scientists for whom science is nothing more than a profession.

Most scientific activity, then, for Kuhn, is puzzle-solving. Among the 'rules' (in an extended sense) which limit the nature of acceptable solutions to puzzles, and the ways in which they can be obtained, are the following:

(I)   Explicit statements of scientific law and about scientific concepts and theories.

(II) Commitments to preferred types of instrumentation and to the ways in which accepted instruments can legitimately be employed.

(III) Higher-level, quasi-metaphysical commitments displayed by historical study.

(IV) Methodological commitments, commitments 'which have held for scientists at all times' (p. 42), such as the commitment to understand the world.

As well as holding over different durations and across different subsets of the relatively disunified body of science, these rules are obviously at very different levels of generality, some concrete and practical, others theoretical. However,

> Though there obviously are rules to which all the practitioners of a scientific speciality adhere at a given time, those rules may not by themselves specify all that the practice of those specialists has in common. Normal science is a highly determined activity, but it need not be entirely determined by rules. That is why . . . I introduced shared paradigms rather than shared rules, assumptions, and points of view as the source of coherence for normal research traditions. Rules, I suggest, derive from paradigms, but paradigms can guide research even in the absence of rules. (p. 42)

This brings us onto the theme of Kuhn's next section, section V, the *priority* of paradigms over rules.

According to Kuhn, historians can identify the paradigms of a mature scientific community pretty easily, by finding a set of recurrent and repeated illustrations of theories 'in their conceptual, observational, and instrumental applications' (p. 43). What they thus find are the tools which members of the scientific community in question use to learn their trade, normally embodied in 'textbooks, lectures, and laboratory exercises' (ibid.).[13] But to determine the paradigms of a mature scientific community isn't yet to identify the shared rules which members of that community follow. The rules followed are *abstracted from* paradigms, and are to be found by comparing the community's paradigms with one another, and with its current research reports. This search for rules, says Kuhn, is 'both more difficult and less satisfying than the search for paradigms' (ibid.). It is frustrating because scientists can agree in their *identification* of a paradigm, without agreeing in their *interpretation* of it. Agreeing in the identification of a paradigm is merely a matter of the scientists in a given field acknowledging that one of their number 'has produced an apparently permanent solution to a group of outstanding problems' (p. 44). A paradigm, though, can guide research in the absence of a shared interpretation.

Kuhn acknowledges that this theme was prefigured by Polanyi, who had argued that science largely depends upon the scientist's '*tacit knowing*', knowing which cannot be articulated explicitly, and thus cannot be communicated via words, but which has to be acquired through practice.[14] Polanyi contrasted the articulable *contents* of science, which one can encapsulate in books, with the 'unspecifiable art of scientific research' (Polanyi 1958, p. 53), which cannot:

> [T]he actual foundations of our scientific beliefs *cannot be asserted at all*. When we accept a certain set of pre-suppositions and use them as our interpretative framework, we may be said to dwell in them as we do in our own body. Their uncritical acceptance for the time being consists in a process of assimilation by which we identify ourselves with them. They are not asserted and cannot be asserted, for assertion can only be made *within* a framework with which we have identified ourselves for the time being; as they are themselves our *ultimate* framework, they are *essentially inarticulable*. (ibid., p. 60, emphasis added)

Polanyi's approach was a necessary corrective to the then-dominant ones, giving Kuhn the idea that (although normal science consists in the articulation of paradigms) paradigms are *not fully articulable*. Kuhn was, in this respect, merely one of a number of mid-twentieth-century philosophers for whom science is as much a matter of skilled *practice* as explicit theory.[15]

Following the well-known discussion of 'family resemblance' concepts in Wittgenstein's *Philosophical Investigations* (Wittgenstein 1958, §66ff.), Kuhn then urges that the various research problems and techniques current within a normal-scientific tradition need have nothing in common, but may be related only by resemblance. Because scientists do not need to know why the models they acquire through their scientific education have the status of paradigms, they do not need any full set of rules which tell them why. Paradigms, that is, 'may be prior to, more binding, and more complete than any set of rules' (p. 46). They guide research primarily by what Kuhn calls 'direct modelling' (p. 47). This is the reason why Kuhn regarded his term 'paradigm' as indispensable, and preferred it to terms urged on him by others, such as 'basic assumption', 'intellectual framework', 'conceptual model', etc. Paradigms, unlike such things, are largely inarticulable. The result is that although historians can *identify* paradigms, they can't *fully* discover or say what paradigms are. Being exemplars, paradigms are the sort of things scientists learn *by example*, rather than by any other kind of linguistically explicit instruction.

This important aspect of Kuhn's views, which commentators often miss or downplay, marks a difference with most preceding philosophies of science, including the logical positivist and logical empiricist tradition, Koyré's intellectualism, and the work of the French conventionalist philosophers (conceived broadly, to include Pierre Duhem). Kuhn's views do have something important in common with those of the conventionalists, in stressing certain 'conventional' components of science. But Kuhn doesn't conceive the range of aspects of paradigms between which scientists really *choose* as widely as conventionalists do.[16] This helps explain what we might call the *invisibility* of paradigms. As David Bohm, one of the first to register Kuhn's focus on these '*tacit*', or '*subliminal*' aspects of science once put it, during normal science scientists 'do not regard its basic character as open to question. Rather, they feel that they are not following a paradigm at all, but are instead simply investigating the actual structure of the world' (Bohm 1964, p. 378). What paradigm-scientists *share* is a *practice*, not a body of theory, rules, definitions, or any other sort of statements. (However, although scientists in a given scientific community share a *paradigm*, they don't share the same *understanding* of it (pp. 44, 50).)

Here there is a further connection with the views of Polanyi, and of Wittgenstein, in the recognition that science has a basis that's neither rational nor irrational, true nor false, but *a*rational. In so far as science has a 'foundation', that is not some set or sets of premises, or even a single method, but rather certain forms of *training*. Scientists fall into communities not because they agree in their beliefs, but because they agree in what Wittgenstein called '*forms of life*', i.e. their training, their education. Among the things inculcated in that training is a set of intuitions about *similarity*. Scientists, that is, are trained to see certain aspects of the phenomena they are studying as similar to others. Kuhn's concept of a paradigm encompasses these similarity relations and, as we shall see, the radical nature of paradigm-changes is largely down to the fact that in such changes, one such set of intuitions about similarity are replaced by another.

In support of this 'priority of paradigms' over rules, Kuhn makes four points. The first is simply the extreme *difficulty* of discovering the rules that guide particular normal-scientific traditions. If rules did guide normal science, they should be relatively easy to recover from the actions and explanations given by the rule-followers in question, i.e. the scientists.

The second point is that because scientists always learn concepts, laws and theories as they're *applied*, and because those applications accompany the theory in question into the textbooks from which future scientists learn their trade, the process of learning a theory depends on the study of applications. There's no *point* in supposing that at any stage in the learning process scientists intuitively abstract a complete set of rules for themselves. Their ability to do successful research can be understood without recourse to 'hypothetical rules of the game' (p. 47). Although paradigms represent a kind of consensus, it isn't a consensus about *statements*, such as axioms, or the definitions of theoretical terms (ET, p. xviii). Scientists aren't taught such definitions, Kuhn claims, but 'standard ways to solve selected problems' in which theoretical terms figure (ET, p. xix, RSS, p. 298).

The third argument concerns the historical pattern of scientific development. According to Kuhn, rules have an important role in that development only during phases when the security of a para-digm is threatened:

The pre-paradigm period, in particular, is regularly marked by frequent and deep debates over legitimate methods, problems and standards of solution, though these serve rather to define schools than to produce agreement. [But] debates like these do not vanish once and for all with the appearance of a paradigm. Though almost non-existent during periods of normal science, they recur regularly just before and during scientific revolutions, the periods when paradigms are first under attack and then subject to change. . . . When scientists disagree about whether the funda-mental problems of their field have been solved, the search for rules gains a function that it does not ordinarily possess. While paradigms remain secure, however, they can function without agreement over rationalization or without any attempted ratio-nalization at all. (pp. 47–9)

And the fourth point is that seeing science in terms of paradigms instead of rules makes the diversity of scientific fields and specialisms easier to understand. Where explicit rules exist, they're usually shared by a broad scientific group, but paradigms, says Kuhn, need not be. Small scientific revolutions, which couldn't be explained in terms of the replacement of explicit rules, can still be explained in terms of paradigm-shifts.

Because it is constituted by its paradigms, normal science as a whole, Kuhn stresses, *isn't* a single monolithic and unified enterprise that stands or falls with any *one* paradigm. Often it seems 'a rather ramshackle structure with little coherence among its various parts' (p. 49, see also Kuhn in Crombie 1963, p. 387). This is an important comment, which suggests that Kuhn thinks the degree to which science is monolithic or unified *changes* over time. The idea that although scientists across a field share a paradigm, it may not amount to the *same* paradigm for them all is also pertinent (p. 50).

In the important postscript to his book, first published in 1970, Kuhn tells us quite a lot more about paradigms, and develops the concept of a scientific community. He worries, first, that the terms 'paradigm' and 'scientific community' are interdefined in SSR, and notes that if he were rewriting the book it would begin with discussion of the community structure of science.

Kuhn starts from what he calls 'the intuitive notion of [scientific] community' (p. 176), as 'the practitioners of a scientific speciality' (p. 177). Within this intuitive notion, scientific communities exist at many different *levels*. At the top, as it were, there is the community of all (natural) scientists. Next there are communities associated with each of the main natural-scientific professions (physicists, chemists, biologists, etc.). At the next level down there are 'major subgroups' such as solid-state physicists, radio astronomers, etc. Membership in communities at these three top levels is easy to establish, but it becomes harder at the next level down. Nevertheless Kuhn is confident that such identification is possible, independently of paradigms, via use of citation data (data about which of their peers' publications any given scientist refers to), for example. Communities at this fourth level, he says, will typically consist of around one hundred members, but individual scientists will often belong to several such groups. It is *these* communities, Kuhn says, which share paradigms.

At this point (p. 179, see also RSS, p. 295) Kuhn admits to another defect of SSR. There, scientific communities are simply identified with scientific subject-matters. But the way we *now* identify scientific subject-matters need not have been the way they were identified in the past. We may now count as 'physicists', for example, scientists who would not have been thought of thus by themselves or by their peers. An important insight of Kuhn's here is that the structure of the disciplines working on a given field of phenomena can itself change (RSS, p. 290). Scientific communities can therefore be identified with scientific subject-matters only *as those subject-matters were conceived at the time in question.*

Having identified such a community, though, one can *then* go on to ask what its members share. SSR's answer, as we know, was 'a paradigm' or 'a set of paradigms'. But in so far as this means an *exemplar* or set of *exemplars*, Kuhn now thinks it inappropriate. Instead, he suggests taking 'paradigm' here in the less important but more popular sense, in which it means 'disciplinary matrix'. As regards this sense, Kuhn tells us that all scientists who work on the same disciplinary matrix will share:

(a) certain *symbolic generalizations* – theoretical assumptions and 'laws' that are deployed 'without question' (e.g. Newton's three laws of motion);
(b) certain *models and analogies*, (e.g. seeing electrical circuits on the model of steady-state hydrodynamical systems);
(c) an idea of what are the good-making qualities of theories, the scientific *virtues* or 'values' (e.g. simplicity, accuracy, consistency, coherence);
(d) certain *metaphysical principles* – untestable assumptions which play a role in determining the direction of research (e.g. the corpuscularian hypothesis);
(e) certain *exemplars* or concrete problem situations, which provide agreement on what constitutes the real *problems* in that field, and on what would constitute their *solution*.

These last components of disciplinary matrices, of course, are paradigms in the *other* sense, exemplars.

Study questions

1. Is a paradigm really a permanent achievement (p. 25), or merely an apparently permanent one (p. 44)? If a new paradigm is also

a promissory note, what assurance do scientists have that it will be able to tackle the puzzles in which normal-scientific activity consists?

2. Should scientists actively seek out anomalies, and regard them as more threatening than Kuhn suggests they do? Is there any alternative to the restricted vision that normal science imposes which would still preserve the potentially revolutionary character of normal science?

3. Should society exercise control over the puzzles or problems which scientists address?

4. Why is it so hard for the scientists involved to identify paradigms, but relatively easy for the historian of science to do so?

5. Would a wider conception of rules succeed in reinstating the idea that scientific activity is a matter of rule-following?

SECTION 4. ANOMALIES, CRISES, AND HOW SCIENTISTS REACT TO THEM

Sections VI and VII of SSR characterize the potential failures of, the fragmentation of, and the exit from, normal science. They tackle the problem: given that normal science does approximately fit the usual image of scientific work as piecemeal accumulation, how does it ever come to result in novelties, either of fact (discoveries) or of theory ('inventions')? Kuhn's answer is that normal science (work under a paradigm) constitutes a uniquely powerful way of challenging and ultimately overthrowing paradigms. The restriction of vision that commitment to paradigms involves is a pre-condition for in-depth empirical investigation, and for the recognition of anomalies. Section VIII explores how scientists respond when anomalies turn, as they sometimes do, into crises. Kuhn here contrasts his view with Popper's falsificationism.

Periods of normal science are characterized by stability and the cumulative extension of knowledge: normal scientific research adds results to the growing stockpile of scientific knowledge. Such periods fit quite well the usual image of science, but they don't include the elements of *discovery* or *invention* that the usual image of science also holds in such esteem. Normal science, Kuhn says, 'does not aim at novelties of fact or theory and, when successful, finds none. New and unsuspected phenomena are, however, repeatedly uncovered by scientific research, and radical new theories have again and again been invented by scientists. History even suggests that the scientific enterprise has developed a uniquely powerful technique for producing surprises of this sort' (p. 52).

The discoveries and inventions in question are aspects of paradigm-*changes*, i.e. scientific *revolutions*. But how does this happen? How does normal science, research under a paradigm, come to constitute an effective way of inducing paradigm *change*? After all, normal science seems a deeply conservative and *un*-revolutionary activity. The general shape of Kuhn's answer is that normal-scientific research, like any other human activity, doesn't produce *only* what its practitioners aim at. Scientific revolutions are results of human action, but *not* of human design: they are *unintended by-products* of intentional activity.

Normal science is characterized by the kind of intensive research which Kuhn calls 'puzzle-solving', and which inevitably turns up *problems*. The normal scientist's job is to make nature fit the existing paradigm, to solve these 'puzzles'. To fail in this task is for the scientist, but not for their paradigm, to fail. When scientists do fail in this task, we have what Kuhn calls an *anomaly*, 'a recognition that nature has somehow violated the paradigm-induced expectations that govern normal science' (pp. 52–3).[17] Anomalies are explored within normal science. Some will be resolved by finding ways to explain the recalcitrant phenomenon. But other anomalies won't be resolved. Some are recognized for many years *without* inducing scientists to reconsider their theories. And some will prove intractable. It is *these* which will ultimately provide the impetus for discovery.

In section VI Kuhn distinguishes, in a loose and informal way, between two sorts of discovery: discoveries of *fact* and discoveries of *theory* ('inventions'). His central thesis here is that discoveries are *processes* or *episodes* rather than *events*. As processes, they have a typical structure, starting with scientists becoming aware of an anomaly, continuing with their exploring the area surrounding the anomaly, and finishing when the paradigm theory has been adjusted in such a way that those phenomena are seen in a different way, what was previously anomalous now becoming expected. Because discoveries of fact are processes, questions such as '*When* was oxygen discovered?' are ill-formed. And because they are processes which involve the mobilization of paradigms, not just individual scientists, questions like '*Who* discovered oxygen?' are also ill-formed. Such questions, the ones the usual image of science prompts historians to address, simply have no answers. Discovery is a process which involves recognizing not just *that* something is but also *what* it is. But recognizing *what* something is involves conceptualization, and concept-formation is something that takes time, sometimes years.

Really fundamental discoveries can bring about, or at least involve, a change in paradigm. Often what is announced as a 'discovery' is in fact the genesis of a new *theory*. Kuhn discusses several examples of scientific discovery (the discovery of oxygen, of X-rays, and the development of the Leyden jar), in an attempt to generalize about its features. In considering how discoveries emerged, he says that 'the perception of anomaly – of a phenomenon, that is, for which his paradigm had not readied the investigator – played an essential role in preparing the way for perception of novelty' (p. 57). The new phenomenon violates deeply entrenched expectations, both theoretical and instrumental.

How it does so is, it has to be said, something of a mystery. Kuhn takes the *categories* of the existing theory to be in outright conflict with those of the new conceptualization, either in that they contradict each other, or in that belief in the old categories implicitly constitutes a commitment to the *non-existence* of items which fall under the new categories. Categories themselves, though, not being claims, cannot contradict one another, and it is unclear why we should think that any associated existence claims are *exclusive* in the way Kuhn thinks. In what way do scientists who come to believe in objects of a certain kind thereby deny that objects of *other* kinds exist? That this isn't *always* the case is clear from Kuhn's own example of Roentgen and his contemporaries, whose paradigms didn't prohibit the existence of X-rays (p. 58). Kuhn appears to think that it's the *scientists* in question who make such exclusivity assumptions, rather than their theories (p. 59). But then the question is why they don't just *revise* their assumption. The general question of when and why theories or paradigms are truly incompatible with one another threatens to raise metaphysical issues which Kuhn doesn't really attempt to deal with.

Kuhn lists the typical characteristics of scientific discoveries as follows:

the previous awareness of anomaly, the gradual and simultaneous emergence of both observational and conceptual recognition, and the consequent change of paradigm categories and procedures often accompanied by resistance. (p. 62)

He then proceeds to argue, by reference to a psychological experiment by the Harvard psychologists Jerome Bruner and Leo Postman, that these characteristics are built into perception itself. By briefly presenting experimental subjects with both normal and 'incongruous' playing cards (black hearts, red spades, etc.), Bruner and Postman showed that

> Perceptual expectancies, whether realistic or wishful, continue to operate so long as they are reinforced by the outcome of events. In short, expectancies continue to mold perceptual organization in a self-sustaining fashion so long as they are confirmed. (Bruner and Postman 1949, p. 208)

In a passage that makes an interesting complement to Popper's normative account of falsification, they argued that

> For as long as possible and by whatever means available, the organism will ward off the perception of the unexpected, those things which do not fit [its] prevailing set. [M]ost people come to depend on a certain constancy in their environment and, save under special conditions, attempt to ward off variations from this state of affairs. (ibid.)

'Perceptual organization', they concluded, 'is powerfully determined by expectations built upon past commerce with the environment' (ibid., p. 222).

For Kuhn, Bruner and Postman's experiment then becomes a model for the process of scientific discovery. The *norm*, both in perception and in scientific research, is the presence of features which are to be expected. Novelties (such as the incongruous playing cards) are perceived only with difficulty, against resistance, within a setting provided by what is expected. At first, only what is anticipated is experienced. In fact, novelties aren't perceived *at all* when they are first presented, and some subjects *never* come to see them. Most people, however, after prolonged exposure to the novelties, become aware of what Bruner and Postman called a 'sense of wrongness'. As a consequence of this feeling, 'conceptual categories are adjusted until the initially anomalous has become the antici-pated' (p. 64). Then, discovery can genuinely be said to have taken place.

But this, according to Kuhn, gives us the answer to our question, 'Why does revolutionary scientific change arise out of normal science?' The negative aspect of this answer is that normal science involves a degree of professionalization which restricts the vision of the scientist, as well as a constitutional resistance to paradigm change. On the more obviously *positive* side, normal science also leads to a wealth of detail, and a precise fit between theory and observation, that 'could be achieved in no other way' (p. 65):

> The areas investigated by normal science are, of course, minis-cule; the enterprise now under discussion has drastically restricted vision. But those restrictions, born from confidence in a paradigm, turn out to be essential to the development of science. By focusing attention upon a small range of relatively esoteric problems, the paradigm forces scientists to investigate some part of nature in a detail and depth that would otherwise be unimaginable. And normal science possesses a built-in mecha-nism that ensures the relaxation of the restrictions that bound research whenever the paradigm from which they derive ceases to function effectively. (p. 24)

These features, rigidity and predictive accuracy, make normal science a taut and sensitive indicator of the presence of problems:

> Without the special apparatus that is constructed mainly for antic-ipated functions, the results that lead ultimately to novelty could not occur. And even when the apparatus exists, novelty ordinarily emerges only for the [person] who, knowing *with precision* what [she or] he should expect, is able to recognize that something has gone wrong. Anomaly appears only against the background pro-vided by the paradigm. The more precise and far-reaching that paradigm is, the more sensitive an indicator it provides of anomaly and hence of an occasion for paradigm change. (p. 65)

Because normal science forms a settled background of precise expectations, novelties stand out clearly. *Without* that background, they would not do so, and scientists would be unable to tell problems from achievements. What Kuhn once characterized as the 'dogmatism' of mature science, the tendency for paradigm scientists tenaciously to defend their paradigm in the face of anomalies, ensures that the most serious of those anomalies spells the downfall of a paradigm only once that paradigm's potential resources for dealing with it are *exhausted.*

Scientists, then, *need* paradigms (exemplars). *Before* they have paradigms what they do constitutes something less than science, and when their paradigms are eventually undermined, their activity is less focused, more philosophical, and largely directed to finding a *new* paradigm. *Mature* sciences, at least, cannot continue without paradigms.

Do these ideas justify the identification (e.g. by Feyerabend) of Kuhn as a *paradigm-monist,* one who thinks that each science, at any given time, can only have a *single* paradigm? This will depend which sense of 'paradigm' we have in mind. Kuhn is more prone to monism about *disciplinary matrices,* than to monism about exemplars. There are certain comments in SSR which reveal that he thinks that at any given time, each community involved in each scientific speciality pays homage to more than one *exemplar.* This is clearest at the beginning of section V, where he talks, for example, of the *paradigms* members of such communities study and practise, and 'the paradigms of a mature scientific community' (p. 43). So when, in the Preface, Kuhn admits there that there *are* circumstances, albeit rare ones, in which two paradigms can peacefully coexist (p. ix), this comment must presumably pertain to disciplinary matrices. It suggests that his view is that mature scientific fields are *usually* dominated by a single disciplinary matrix. (Whether scientific fields can be identified *independently* of disciplinary matrices, though, is a good question.)

But perhaps Kuhn's work does contain materials for a certain kind of pluralist critique of science. Science, if Kuhn is right, works with an institutional mechanism which assumes that the natural world can be only one way. (That is what the presumption that rival paradigms *exclude* one another amounts to.) Kuhn, though, was groping towards a perspective that suggests how this might not be the case, how genuinely different but nevertheless equally good conceptual schemes might apply to the same phenomena (see section 6 of this *Guide,* below). If this is right, science has institutionalized a mechanism that prohibits a perfectly legitimate *pluralism* about conceptual schemes (or 'disciplinary matrices').

So much for novelties of fact. What about novelties of theory, which Kuhn calls 'inventions'? These, he argues, are the more important factor in the production of paradigm-shifts. Discoveries of fact are never sufficient to produce a change in paradigm. They must be accompanied by theoretical innovation. But now we face a question parallel to the question about factual discoveries: how do *theoretical* innovations arise from normal science? The problem is exacerbated by the fact that normal science seems to be even *less* directed towards theoretical innovation than towards factual discovery!

Kuhn's answer is very largely parallel to his answer to that previous question. Awareness of anomaly again plays a leading role. The history of the natural sciences shows that before the transition to a new paradigm the old paradigm is always perceived to be in trouble in some way. The trouble usually takes the form of the failure of normal science puzzles to turn out as the paradigm says they ought. This can multiply until what we have is a breakdown in puzzle-solving activity, which then constitutes the core of what Kuhn calls *'crisis'*. External factors such as political or religious conditions, and the personal circumstances of particular scientists, etc. can determine *when* a crisis occurs, when it comes to be recognized, and even perhaps the area of theory in which the breakdown happens, but they can't precipitate crisis itself. That's a matter *internal* to the paradigm.

Here we can contrast Kuhn both with his most illustrious forebear, and with certain more determinedly sociological thinkers. Koyré had already made a breach in the idea that the development of science was *autonomous*, but only to the extent of showing its relations to other intellectual disciplines: philosophy (notably, metaphysics) and religion. By understanding the particular scientific ideas he studied in their purely intellectual context, and relating them so closely to philosophy, Koyré was widely perceived to have risked 'disembodying' them from their *social* and *technological* context. Other historicists and sociologists of knowledge went much further, wanting to show that the development of science depends decisively on precisely such 'external' influences. Kuhn, too, sought to go beyond Koyré in this respect, but *only* to a rather limited extent.

While he did pay attention to the social context of scientific activity, Kuhn shouldn't be thought of as having done more than *introduce* discussion of social factors to Anglo-American philosophy of science; he was an 'internalist' rather than an 'externalist' (see Kuhn's 1968 article 'The History of Science', in ET, Hacking 1979, p. 225, and RSS, pp. 287–8). By his own admission, he ignored the role of technological and social conditions, thinking that these would not affect SSR's main theses (p. x). He insisted, rightly, that whether or not social contextual features are relevant to any area of inquiry *must* be an empirical and therefore open question. But he then went on to argue that the special *social* nature of scientific communities, their isolation and insulation from 'external' factors, was the reason why one generally didn't have to take account of such factors in understanding the development of scientific ideas. Mature science, he felt, was 'more fully, though by no means completely, insulated from its social milieu' than any other discipline (ET, p. xv). So while he thought of his own work as 'deeply sociological' in a certain way (ET, p. xx), since it portrayed science as the product of *groups*, not of individuals, he also inveighed against certain sociologists of science who presented themselves as 'Kuhnians', accusing them, for example, of seriously underestimating the role and significance of *common values* in science (ET, pp. xxi–xxii).

One of the signs of crisis is the re-emergence of some of the features which characterize the pre-paradigm period, such as the *proliferation of versions*. In extreme cases there are as many versions of the theory as there are leading scientists working on it. Foundational issues are explored in the hope that they will yield new fundamental insight. The solid core of agreement which characterizes normal science is eroded away. This makes it harder and harder to see what the theory *is*.

Kuhn mobilizes three examples of crisis (that which preceded the emergence of Copernican astronomy, that which preceded the oxygen theory of combustion, and the crisis in late-nineteenth-century physics) which he considers typical. In each of them, new theories emerged only after a crisis preceded by a distinct failure in normal-scientific puzzle-solving, and as a direct response to that crisis. But, ironically, the problems which precipitated the breakdown had been around for some time, and had been considered already *solved* under the existing paradigm. The breakdown, therefore, showed that even on its home ground, a paradigm cannot ensure that its solutions to problems are *definitive*. Paradigms can be embarrassed by having their own most illustrious successes collapse on them. Kuhn suggests that the new solutions to the problems in question may well have been anticipated by the development of new theories during the preceding period of normal science. But his functional perspective leads him to insist (against Feyerabend, for example) that the development of such theories, being speculative, represents an extravagance at that time. A critic like Feyerabend would of course complain that if scientists *didn't* come up with new theories, even during 'normal' science, crises might be even *less* prevalent than they are. But perhaps Kuhn might reply that although crises are a necessary condition for the 'emergence' of new theories (p. 77), meaning their *advocacy*, they aren't, strictly speaking, a necessary condition for their *existence*. Paradigm scientists, that is, are free to invent new theories, but they are not free to advocate them until something has gone seriously wrong with their existing theory.

How, then, do scientists respond to crises? If one followed the Popperian model, and identified what Kuhn calls 'anomalies' with what Popper calls 'falsifying instances', one would be led to believe that they respond by giving up their theory. Kuhn instead claims that, as a matter of historical fact, scientists *don't* respond to crisis by giving up their *paradigm* (even when that paradigm *is* a theory). They don't compare their paradigms directly with nature, and they don't treat 'anomalies' as falsifying instances, even though 'in the vocabulary of philosophy of science that is what they are' (p. 77).

Kuhn's account of paradigm-transition explains why scientists don't abandon their paradigm, even when it has encountered problems. An accumulation of anomalies is a necessary but not sufficient condition for such a move. On Kuhn's account, abandoning a paradigm must *always* involve taking up an alternative paradigm. Scientists only switch allegiance when there's some more attractive paradigm to switch allegiance *to*. This is because science *without* paradigms is inconceivable. For a scientist to reject a paradigm simply because of anomalies (counterinstances) is to *cease to be a scientist*, to leave the scientific community.

The way Kuhn has identified paradigms, after all, means that they include, at least, the scientist's conceptual 'toolkits', their concepts, and without some such toolkit, a scientist is unemployable (see Kuhn in Crombie 1963, p. 387). Kuhn's story also emphasizes that the rejection of a paradigm doesn't arise just from comparing it with 'the facts'. 'The decision to reject one paradigm is always simultaneously the decision to accept another, and the judgment leading to that decision involves the comparison of both paradigms with nature *and* with each other' (p. 77).

It's tempting to compare Kuhn's views with Popper's here (as Kuhn himself does), and declare that Popper's story fails to explain both why scientists with problematic theories don't just drop them, and why scientists compare theories with one another, as well as with the results of observation. This, though, would be misleading, if not plain wrong. First, the theories Popper was talking about aren't always the paradigms that Kuhn was talking about. One can only directly compare their views where the paradigms in question *are* theories. Even then, Kuhn's interest in the actual sociological fate of those who relinquish their paradigm theory would not quite match up with Popper's interest, which is in how the people in question *ought* to be treated. Second, Popper *does*, right from the start, allow a role for theory-comparison in theory-testing. He insists that a theory is tested for internal logical consistency, by way of empirical applications of its conclusions, and by 'comparison with other theories, chiefly with the aim of determining whether the theory would constitute a scientific advance' (Popper 1959, pp. 32–3). One might still complain, though, that Popper didn't make much of this role.

It's just as well that the presence of counterinstances doesn't cause scientists to abandon their paradigm, since paradigms, Kuhn insists, are *always* confronted with counterinstances. In fact, normal science *consists* in the work scientists do as a response to these. Scientific theories which really do solve all their problems cease to be parts of pure science and become engineering tools. In a sort of Gestalt-switch relationship, one person's 'puzzles' are another's 'counterinstances'.[18] Without these puzzles/counterinstances, there would *be* no normal science. But scientists spend most of their time *applying* their theories, not *testing* them by looking for negative evidence. That's why Kuhn feels justified in treating the scientific work in question as the treatment of puzzles *rather than* as responses to counterinstances. Scientists *do* engage in theory-testing, as well as in other activities which fit the stereotype of scientific activity, such as constructing speculative theories and trying speculative experiments, but they do so, according to Kuhn, principally during periods of *extra*ordinary science, after one or more significant anomalies have already set in.

When considering how scientists respond to anomalies, then, it's important to note that even 'persistent and recognized anomaly does not always induce crisis' (p. 81). The reason is that anomalies don't always *remain* anomalies: some of them are *removed* by subsequent puzzle-solving activity. If we now ask, 'What makes an anomaly worthy of serious scrutiny?', Kuhn tells us that there's no general answer (and his account of when puzzles turn into crises is perhaps the sketchiest part of the book). The kinds of factors which enter the process are that the anomaly might be seen as calling into question the paradigm's fundamental generalizations, or that it inhibits applications with a particular practical importance. When factors like these conspire together, the anomaly becomes more than just another puzzle within normal science. It becomes generally recognized as a real problem. If it resists the application of concerted activity on the part of the paradigm's leading scientists, its resolution becomes a point of honour. The activity that is then focused upon it causes minor deviations from the paradigm. Intense research goes into finding out what kinds of small alterations to the paradigm will yield a solution. The existence of minor deviations becomes exacerbated into the phenomenon of competing 'schools', and as a result, the paradigm itself is blurred. The solid agreement which constitutes the shared paradigm begins to erode, and attempts are made to articulate alternative theoretical structures. During this period, scientists will even go so far as to resort to philosophical analysis in order to find lines of solution. They will begin to examine the philosophical assumptions on which their existing paradigm is based, and doubts about these assumptions will be expressed. This is when the phenomenon of 'thought-experiments' comes into its own. It marks the onset of another period of non-normal science, a period of *revolutionary* scientific change, which is *'a reconstruction of the field from new fundamentals'* (p. 85, emphasis added).

Kuhn tells us that 'By concentrating scientific attention upon a narrow area of trouble and by preparing the scientific mind to recognize experimental anomalies for what they are, crisis often proliferates new discoveries' (p. 88). How does this happen? Sometimes the new paradigm is foreshadowed in the extraordinary research that has just gone into solving the anomaly. In other words, sometimes new paradigms *are* the result of solving some old problem. At other times, 'no such structure is consciously seen in advance. Instead, the new paradigm, or a sufficient hint to permit later articulation, emerges all at once . . . in the mind of a [scientist] deeply immersed in crisis' (pp. 89–90). This individual creativeness Kuhn treats as ineffable.

His sociological approach allows him to note that it's usually a *young* scientist who has the new idea, since the young are less tightly bound into the old paradigm, but it doesn't allow him to explore the 'inscrutable' (ibid.) nature of the psychological process itself.

In his Postscript (p. 181), Kuhn queries whether crises precede revolutions as invariably as SSR implies. But he argues that nothing essential hangs on the answer, that crises need only be the *usual* preludes to revolution, and that crises may be generated, not by the work of the paradigm-scientists in question, but by the work of scientists in related areas.

Study questions

1. Does Kuhn successfully explain how conservative activities can have revolutionary results, how 'the productive scientist must be a traditionalist who enjoys playing intricate games by pre-established rules *in order to be* an innovator who discovers new rules and new pieces with which to play them' (ET, p. 237, emphasis added)?
2. Why should the historian's inability to answer a question (such as 'When was oxygen discovered?) mean that such a question is intrinsically problematic?
3. Are 'anomalies' merely in the eye of the beholder? Are Kuhn's examples enough to convince us that scientific discoveries are always preceded by the awareness of anomaly?
4. If systems of categories differ, why should they be thought of as incompatible? Why should one such system be thought of as denying the existence of items falling under the categories of another such system?
5. If scientists didn't have their vision restricted by their commitment to paradigms, would they be able to perceive anomalies? Is an informed mind better than an open mind in this respect? Is the restriction involved in commitment to a paradigm a necessary pre-condition for any in-depth investigation of nature?
6. Would scientists be right in responding to a crisis by giving up their paradigm? Would they be right to respond by giving up their *theory*? Does Kuhn's view of theory-testing have advantages over Popper's? Is science without paradigms possible? Is there reason to think such an activity would make significantly less epistemic progress than actually-existing science? In the absence of puzzles, would 'normal science' cease to exist?

SECTION 5: SCIENTIFIC REVOLUTIONS VERSUS CUMULATIVISM

SSR section IX raises questions about the nature of scientific revolutions, and their function in the development of science. Kuhn seeks to justify his central political metaphor: the rest of his book is supposed to demonstrate that the historical study of paradigm-change reveals parallels between science and politics. In this section, Kuhn sets up a conflict between the resulting view of science and the usual, cumulativist view, and suggests reasons for doubting the latter.

What of scientific revolutions? Kuhn has already told us that

The transition from a paradigm in crisis to a new one from which a new tradition of normal science can emerge is far from a cumulative process, one achieved by an articulation or extension of the old paradigm. Rather it is a reconstruction of the field from new fundamentals, a reconstruction that changes some of the field's most elementary theoretical generalizations as well as many of its paradigm methods and applications. (pp. 84–5)

One of the central elements in such a transition, he argued in section VIII, is a change in the *perceptions* of the members of the scientific community. The scientists will have changed their perceptions of the field, its methods and its goals. Here Kuhn invokes the controversial 'Gestalt-switch' metaphor, which we shall return to in the next section.

Another metaphor that Kuhn is using is apparent in the title of his book itself, the *political* metaphor of revolution. He argues that the parallels between political revolutions and paradigm transitions are rich enough to support this metaphor. Some aspects of these parallels are (pp. 92–3): the ways in which both kinds of revolutions come into being through an awareness of the inadequacy of the existing institutions, an awareness that they are failing to solve problems; the idea that in both the sense of malfunction is a prerequisite to revolution; the fact that in each area one is faced with 'a choice between *incompatible* modes of community life' (p. 94, emphasis added); the idea that revolutions aim to change the existing institutions in ways which those institutions themselves prohibit; and the idea that in the interim transition period normal canons of debate are suspended in favour of decision-making methods involving techniques of persuasion rather than argument. As we shall see, this last point is one of the most important.

Scientific revolutions, then, are 'those non-cumulative developmental episodes in which an older paradigm is *replaced* in whole or in part by an *incompatible* one' (p. 92, emphases added). The same applies to the invention of new theories (p. 97). But why must the acceptance of a new paradigm or theory mean the *rejection* of the existing one?

One can imagine new paradigms and theories being introduced which don't disturb the old ones, but according to Kuhn this rarely or never actually happens. This brings him into conflict with philosophers of science, such as the logical empiricists, who envisaged science growing cumulatively through the development of new theories because they believed that old theories were *reducible*, by logical derivation, to new ones.

Kuhn associates cumulativism with a particular *epistemological* tradition or 'paradigm' which has dominated Western philosophy since the scientific revolution. (He tells us more about this in section X.) But, as we already know, he feels that there is increasing reason to doubt the latter. Some of these reasons derive from psychology, but Kuhn also suggests that cumulativism and its associated paradigm will be undermined if only we take the history of science seriously. He takes it that an unbiased look at that history will inevitably reveal the existence of a multitude of scientific revolutions.

This, it has to be said, is too superficial. Cumulativists are too many and probably too various to be clearly tied to a single philosophical 'paradigm'. The logical positivists and logical empiricists may be open to the objection that their image of science was formed not by attending to actual science and its history, but rather by working under the *a priori* presumption that scientists must use methods which make sense when represented in formal logic. But cumulativism doesn't arise only from 'logical', as opposed to historical, approaches to science. It's naive to think that an unbiased look at the history of science *must* result in non-cumulativism.

Cumulativist history has been written not only by 'practitioner' historians (i.e. scientists who write history) but also by professional historians and philosophers of science, including figures such as William Whewell, Émile Meyerson, Kuhn's mentor James Bryant Conant and George Sarton.

The prominent nineteenth-century scientist, historian and philosopher William Whewell, just to take one example, is a good example of a sophisticated cumulativist historian of science. He and Kuhn share approaches and views at least as important as those that divide them, including a thorough-going historicist insistence on seeing science as formed by processes which develop in time, a recognition that science proliferates and progresses by division and subdivision, and an assessment of metaphysics as essential to science. Whewell, like Kuhn, also divided the history of science into 'epochs' of different kinds, including 'stationary periods', and saw great individual scientists as synthesizing elements of a pre-existing context into generalized frameworks. But where Kuhn sees discontinuity in this process, Whewell tried explicitly to go beneath the apparently discontinuous surface of historical events in order to discern the rational developmental relations which drive them. The history of each science, according to Whewell, 'which may . . . appear like a succession of revolutions, is, in reality, a series of developments' (Whewell 1984, p. 8). This method of 'rational reconstruction' was later taken up by the logical positivists, logical empiricists and Imre Lakatos.

In addition, but crucially, there is an entire tradition of Darwinian historians and philosophers of science (including Ernst Mach and Pierre Duhem, but more recently Stephen Toulmin and David Hull) that not only provide an *alternative* to Kuhn's revolutionism, but do so by drawing on a source (evolutionary theory) to which, as we shall see, Kuhn linked his own work.

Kuhn treats Whewell, Mach and Duhem as representatives of a single and determinedly *philosophical* tradition in the historiography of science (ET, pp. 106–7). But his accusation that cumulativists don't take the history of science seriously fails to register the interpretive latitude available when doing history of science. Neither continuity nor revolution is written on the face of science, and to suppose otherwise is to fail to take account of the fact (of which Kuhn was elsewhere well aware) that history is an interpretive (and therefore partly philosophical) discipline.

This is not to say that Kuhn's critique of the *particular* version of cumulativism he discusses is unsuccessful. (Cumulativism can, of course, be naive. But so can revolutionism.) The version in question, which Kuhn finds in early logical positivism, and which is associated with the 'instrumentalist' view of theories, certainly seems restrictive. It says, roughly, that new theories serve not to *overthrow* old ones, but merely to limit their domain of application. The success of Einstein's theory of relativity, for example, rather than showing that Newton's theory fails to apply at all, was supposed to show that it applies only to non-relativistic phenomena.

Kuhn, like Popper, objects that this would render theories *permanently* immune to attack. He suggests, first, that restricting the field of application of theories will also mean that scientists would be prohibited from theorizing about any phenomena which have not yet been observed, and that this would put an end to the research through which science develops. But this objection seems superficial: scientists will still have a legitimate interest in finding the limits within which their theories apply, and as-yet-unobserved phenomena will still be of potential interest in this process. The tentative application of a theory or paradigm to new areas is by no means ruled out.

Along the same sort of lines, though, Kuhn's second and more promising objection is that such restrictions would disable 'the mechanism that tells the scientific community what problems may lead to fundamental change' (p. 101). The normal-scientific commitment is a commitment to apply a particular paradigm not only where it has already been applied, but also to new phenomena, and to apply it with a hitherto unattempted precision. These extensions are what supply a continuing source of *new* puzzles for normal science. Since normal science consists of puzzle-solving, such science could not exist in the absence of such commitment. Likewise, anomalies and crises, the wellsprings of extraordinary science, only arise because of this same commitment to apply paradigms well beyond the competence of their initial applications. The price of scientific progress is the persistent possibility of being wrong. This thought runs against the general empiricist tendency, which positivists share, to seek epistemic *security* in science, by portraying it as keeping as close as possible to what is already known, and to what can be observed.

Kuhn also argues that reduction by logical derivation is an unrealistic goal for the particular case of Newtonian dynamics and relativistic dynamics. The reason why reduction is unrealistic is that the *meanings* of the theoretical terms of the two theories are not the same. Even though both Newtonian dynamics and relativistic dynamics use the term 'mass', they don't mean the same by that term. For Newton, mass is a property, for Einstein, a relation. In fact, in relativistic dynamics there are *two* concepts: proper mass and rest mass. What's more, we are later told that even the meanings of *observation* terms, like 'planet' (p. 128) and 'earth' (p. 149) change, according to Kuhn. A change in the meaning of familiar terms like these is, according to Kuhn, one of the concomitants of scientific revolution. Kuhn later came to express this thesis in rather different and more refined terms. He allied it with his earlier thesis that there is a change in the *ontology* of the theory, arguing that 'the distinctive character of revolutionary change in language is that it alters not only the criteria by which terms attach to nature but also, massively, the set of objects or situations to which those terms attach' (RSS, pp. 29–30).

To the usual story that later theories incorporate earlier ones, Kuhn opposes his view that earlier and later paradigms differ with one another in such a way as to *conflict*. Whether he explains these differences in terms which really make sense of the paradigms conflicting is a troubled issue, though.

Successive paradigms differ in several ways. The first and most obvious, which Kuhn calls 'substantive' differences, is that they tell us different things about the population of the universe. They involve different *ontologies*, for example, different lists of the kinds of objects the world contains, as well as different views about what those objects get up to. But a second way successive paradigms differ arises from the fact that they aren't merely *theories*. Paradigms, as we have seen, also involve conceptions of correct method, of what counts as a problem, and of what counts as its correct solution. When paradigms change, all these things may change with them. Scientific revolutions, therefore, (presumably Kuhn here has in mind the larger kind of revolution, change in disciplinary matrix, rather than mere change in exemplar) involve a reconceptualization of the *science* in question itself, 'a reconstruction of the field from new fundamentals' (p. 85), not just a change in theory. And at this point Kuhn makes his first real mention of the possibility that pre- and post-revolutionary normal-scientific traditions should be '*incommensurable*', rather than merely incompatible (p. 103). (We will deal with this key concept in section 8.)

Kuhn first illustrates these 'subtler' effects of paradigm-shift with the example of the Newtonian revolution (pp. 103–6). He then goes on to suggest that there are other historical cases in which problems solved by old theories are *not* solved by their more recent and better successors. In recognition of his having drawn attention to this phenomenon, it has actually come to be called '*Kuhn-loss*'. It is first illustrated on p. 107, in terms of two examples. In the first, Kuhn explains that the chemical revolution (which involved a *net* gain in explanatory power) involved a partial *loss* of explanatory power in that likenesses between metals, and between acids, could no longer be explained after Lavoisier had rendered the phlogiston theory, with its reference to chemical 'principles', obsolete. In the second example, the acceptance of Maxwell's electromagnetic theory, which came with no good account of a medium in which the propagation of light waves could be explained, is deemed to have involved a loss of explanatory power relative to previous wave theories of light. Later, in section XII, Kuhn adduces a further example of a lost *question*, if not a lost solution: Newton's theory of motion, unlike its predecessors and its successor, famously did not attempt to explain the *cause* of gravitational attraction (p. 148). If Kuhn is right about this, it shows that questions can not only be lost, but also that they can be later rediscovered, and answered.

Kuhn then pauses to argue that the kind of changes in problems and standards which scientific revolutions involve *cannot* be construed as a change from a methodologically lower to a higher type (p. 108). I think he is mistaken in arguing that if one *were* to grant this, one would have succumbed to cumulativism. It is perfectly possible, I suspect, that although there might be no accumulation of scientific *results*, scientific *methodology* continues to improve. (It's hard to see any sense in which methodology could *accumulate*, strictly speaking.) Kuhn declares that it's even harder to argue the case for the latter than the case for the former, but his examples hardly establish this. To defend the idea that methodology generally improves over time, one need not deny the possibility of Kuhn-loss. That is, one can quite well recognize that later scientists sometimes refuse to apply their theories to each and every phenomenon to which previous theories had been applied (as in Kuhn's examples of gravity, and of the colour and aggregation of substances). One can equally well acknowledge that such problems may once again be addressed by even later theories (e.g. Einstein's theory, in the case of gravity). And finally, one need not even deny that methodology *occasionally* degenerates (although Kuhn gives no examples of this). Kuhn, I think it has to be said, takes on a far too simple version of the idea that methodology progresses. In fact, I suspect that a more sophisticated version of that idea is quite compatible with the things he wants to say about scientific progress (see section 8).

From the examples he has given, Kuhn then draws the conclusion that paradigms constitute science in that theory, methods and standards are acquired together, and 'usually' (p. 109) change together. For an idea with such momentous consequences, the few examples given are again insufficient to support such a thesis, and Kuhn does little more to support it. But the thesis represents, I think, an important instance of an over-zealous presumption of *holism* on Kuhn's part. If theory, method and standards do all regularly change together, and if, as Kuhn has argued, method and standards *merely* change, rather than improving, then the comparison of earlier and later paradigms is of course rendered difficult, if not impossible. For we can then say nothing like 'The new theory is more successful than the old one', let alone 'The new theory is more successful than the old one at tackling the same set of problems'. This predicament, arising from such a thorough-going holism, is one of the reasons why Kuhn was accused of *relativism* by certain other philosophers of science. (In sections 6 and 8 we shall see what can be said in his defence on this score.)

Alasdair MacIntyre has taken Kuhn to task on this issue. Kuhn's account of paradigm-change, he complained, assumes 'not just that the adherents of rival paradigms disagree, but that *every* relevant area of rationality is invaded by that disagreement' (MacIntyre 1977, p. 466, emphasis added). Ironically, given his own attempt to distance himself from the prevailing 'Cartesian' philosophical world-picture, Kuhn thus understands what MacIntyre has appropriately called 'epistemological crises' in a way reminiscent of René Descartes' attempt to doubt everything simultaneously. Were this possible, it really would raise the worry that scientific conversion could be nothing more than a blind leap of faith.

And here, of all the places in SSR, is where Kuhn does indeed seem closest to implying that earlier and later scientific products simply cannot be compared in any meaningful way. In so far as normal-scientific traditions disagree about what counts as a problem and what counts as a solution, they will, he says, 'inevitably talk through each other when debating the relative merits of their respective paradigms' (p. 109). The arguments that result will regularly be 'partially circular' (ibid.), since each tradition will live up only to the standards it sets itself, rather than to those set by its opponent. And this 'incompleteness of logical contact' (p. 110) is partly a result of the fact that proponents of rival paradigms will have to consider, and may well disagree about, an issue concerning *values*, namely, which problems it is more *important* to have solved. Under these circumstances, comparing paradigms will not even be like comparing apples with oranges but, even less promisingly, rather like comparing apples with television sets. In turn, this will only exacerbate another problem: the less complete the 'logical contact' between successive paradigms, the harder it will become to see why paradigms should truly *conflict* with one another.

Kuhn, both in SSR and its Postscript, specifically notes his intention to apply the concept of a scientific revolution both to small- *and* large-scale phenomena, since he regarded the possibility of relating the structure of small revolutions to that of large ones (such as the Copernican revolution) as important (pp. 7–8, 180–1). He lists as the most obvious *major* scientific revolutions those associated with Copernicus, Newton, Lavoisier and Einstein (p. 6). The revolution in chemistry consequent upon the work of Lavoisier and Dalton is declared to be 'perhaps our fullest example of a scientific revolution' (p. 133). The transformation of biology that followed Darwin's work is sometimes mentioned as another example (p. 180 and ET, p. 226). The transition from the corpuscle theory of light to the wave theory apparently counts as a revolution (p. 102), and although Kuhn doesn't explicitly say whether it is large or small, the fact that he groups it with the ones mentioned above suggests the former.

SSR's examples of *smaller* revolutions are surprisingly thin on the ground: Maxwell's equations (p. 7), the discovery of oxygen, the discovery of X-rays (pp. 92–3), and perhaps the transition from the caloric theory of heat to the theory of energy conservation (pp. 97–8).

Applying the concept of scientific revolution (and paradigm) to phenomena across these very different scales, it's now generally recognized, was a mistake. As Hacking has argued, by doing so Kuhn deprives himself of anything general to say about *all* paradigms, or *all* scientific revolutions (Hacking 1979, pp. 230–1). Many of the thoughts that make Kuhn's story sound either exciting or dangerous (e.g. world-changes, incommensurability) are prompted by thoughts about *big* scientific revolutions. But there are things to be said about these changes to disciplinary matrices which seem quite different from what can be said about the more everyday changes to exemplars. Maybe there are no 'shared elements that account for the relatively unproblematic character of professional communication and for the relative unanimity of professional judgement' (ibid., p. 233)? Maybe there is nothing in common among the answers to Kuhn's key question 'for different identifiable scientific communities, in quite different times and societies' (ibid.)? And consequently maybe there's no such thing as *the* structure of scientific revolutions (big and small)? What happens in the big revolutions is of a quite different order from finding new exemplars:

On the one hand there is the disciplinary matrix dominating one hundred souls with its acknowledged achievements, its institutionalised hierarchy, and the standard examples taught to students. On the other hand there is 'the new texture of human experience in a new age'. *The Structure of Scientific Revolutions* too easily rides the rollercoaster of a continuum between the two. (ibid., p. 234)

Having now seen the main elements of Kuhn's new image of science, we can pay some attention to the vexed issue of the *status* of his claims. Most commentators take Kuhn to be putting forward some kind of theory. But was he trying to say something *empirical*, or something *conceptual*? Norwood Russell Hanson, one of the figures whose work on the philosophy of science Kuhn was trying to draw upon in SSR, was only one of the people who pleaded with Kuhn to make clear the status of his claims (see Hanson 1965).

Almost all commentators have taken Kuhn to be putting forward empirical claims that could, in principle, be confirmed or refuted by evidence from the history of science (though not necessarily in as simple a way as pre-existing philosophies of science sometimes suggest). Some have even questioned whether *any* events in the history of science really fit Kuhn's image, and this certainly goes towards thinking of his claims as empirical.

There is, however, another possible reading of Kuhn's work. Commentators such as Wes Sharrock and Rupert Read have read Kuhn as trying to wean us away from an existing image of science by showing us that its claims are literally nonsensical (rather than false), and then going on to substitute a picture that doesn't fall foul of our concepts of science, its components and related activities.

This Wittgensteinian interpretation of Kuhn was pre-figured by Mary Hesse, who began one of the very first reviews of SSR by saying 'It is the kind of book one closes with the feeling that once it has all been said, all that has been said is obvious, because the author has assembled from various quarters truisms which previously did not quite fit and exhibited them in a new pattern in terms of which our whole image of science is transformed' (Hesse 1963, p. 286). Hanson, too, considered the possibility that Kuhn's picture of science was not *supposed* to be vulnerable to empirical counter-evidence, worrying that the terms 'paradigm' and 'scientific revolution' were intertwined in such a way as to make the identification of the one depend on the identification of the other.

I don't think Kuhn had a consistent attitude towards this question. He *usually* treated SSR as if evidence from the history of science was relevant to its assessment. In the Postscript to SSR, for example, he expressed concern that the terms 'paradigm' and 'scientific community' were interdefined there (p. 176), and he took pains to ensure that scientific communities could be identified independently of paradigms. He also admitted that SSR's 'theory about the nature of science', like any other theory, *might* be wrong, but that it should be taken seriously because scientists *do* behave as it says they should (pp. 207, 208). Certainly his attitude towards other, *competing* views about the nature of science was that they could be infirmed by comparison with the historical record.

Elsewhere, however, he was more elusive. When queried in 1995 about the relation between SSR and his 1978 book *Black-Body Theory and the Quantum Discontinuity*, he declared that 'you cannot do history *trying* to document, or to explore, or to apply a point of view that is as schematic [as SSR]', and that 'If you have a theory you want to confirm, you *can* go and do history so it confirms it, and so forth; it's just not the thing to do' (RSS, p. 313; see also Sigurdsson 1990, p. 23). John Heilbron reports Kuhn making similar but more obscure comments when explaining why he hadn't used SSR's famous terms 'paradigm', 'revolution', etc. in the later book. SSR, Kuhn said, was 'two times removed from a claim to fit all historical episodes: it was a candidate paradigm, not a covering theory; moreover, even if it were a theory, applying philosophical theories in the writing of history "is just not the thing to do"'. When asked what historical episodes he would assign to doctoral students eager to confirm his theory of revolutions, Kuhn replied, "It is not a theory, and I do not expect it to match the record"' (Heilbron 1998, p. 511).

The idea that one shouldn't 'do history' with the aim of confirming or refuting a theory is well-taken. But this doesn't speak to the issue of whether, after the history has been done, theories can and should be compared against it. That a theory is 'schematic' suggests only that it would need to be fleshed out in order for the comparison to take place. I am not certain what Kuhn can have meant by calling SSR a 'candidate paradigm' rather than a covering theory, but by explicitly *denying* that it was a theory, he certainly seems to have come close to the Wittgensteinian take.

Study questions

1. Are the parallels between science and politics rich enough to justify Kuhn's political metaphor of revolution?
2. What link is there between cumulativism and the dominant epistemological 'paradigm' Kuhn detects? Can there really be paradigms in *philosophy*, and if so, which sense of 'paradigm' is at issue? How might cumulativists respond to Kuhn's critique?
3. What is lost in cases of 'Kuhn-loss'? Can Kuhn-loss be presented as emancipation from problems that are *spurious*?
4. Does Kuhn successfully explain how scientific paradigms conflict with one another? Does logical compatibility make conflict easier, or harder to explain? Why can't paradigm-debates be solved by reference to those rules 'which have held for scientists at all times' (p. 42)?
5. Should we think of Kuhn as making empirical claims about science, or as doing something more philosophical? If the latter, is he propounding a philosophical theory, or rather trying to wean us away from such a theory? If the latter, what does Kuhn put in place of the theory in question?

SECTION 6: GESTALT-SWITCHES AND WORLD-CHANGES

Section X explores how paradigm-changes (changes in disciplinary matrix) should be conceived. Kuhn endorses the idea that they involve world-changes, and explains this in terms of the 'prototype' of visual Gestalt-switches. This section is an exercise in stretching concepts, particularly perceptual concepts.

The historian, section X of SSR begins, is tempted to suggest that when a paradigm-shift occurs, the world changes. Kuhn apologizes again and again for using the idea that paradigm-shifts, that is, changes in disciplinary matrix, are changes in the world itself. He also qualifies it in various ways. But in the end he just can't help himself. 'In a sense that I am unable to explicate further,' he later says,

> the proponents of competing paradigms practice their trades in different worlds. One contains constrained bodies that fall slowly, the other pendulums that repeat their motions again and again. . . . Practicing in different worlds, the two groups of scientists see different things when they look from the same point in the same direction. Again, that is not to say that they can see anything they please. Both are looking at the world, and what they look at has not changed. But in some areas they see different things, and they see them in different relations one to the other. (p. 150)[19]

This passage embodies two of SSR's most controversial ideas, that of *Gestalt-switches*, and that of *world-changes*.

Kuhn uses the idea of a Gestalt-switch as a metaphor for what happens in science education, as well as for what happens in full-scale paradigm-change. Transformations like the Gestalt-shift, he suggests, although *more gradual* and *irreversible*, are common concomitants of scientific training and paradigm-change. At times of revolution, in particular, when the tradition changes, the scientist's perception of the environment must be re-educated. He or she must learn to see a new Gestalt, as a result of which '[w]hat were ducks in the scientist's world before the revolution are rabbits afterwards' (p. 111).

Note that, contrary to what the title of his section X suggests, Kuhn really does mean that what *were* ducks *are* now rabbits, not merely that what looked like ducks now look like rabbits. It is not only a change in 'world *view*' that is in question. Describing things in Kuhn's way brings us to his famous idea of '*world-changes*', which has raised the worry that he subscribes to a kind of philosophical *idealism*. Kuhn, as Hacking says, 'does imply that to have a radically different paradigm . . . is to live in another world. It is not just to describe the world differently (as today's philosophical 'realist' would put it) but, Kuhn implies, to be in another world, and that sounds as if the world in which we live is in part a product of our mental activity' (Hacking 1979, p. 229).

Do we, however, *need* to describe things in this way? There is, of course, a more familiar and apparently less troubling story, which Kuhn opposes, according to which when scientists change their minds about what they observe, they do so merely by *reinterpreting* stable observational data, 'sense-data'. Kuhn associates this story with the Western epistemological picture or 'paradigm' we mentioned in the previous section.[20] Science, according to this paradigm, is 'a construction placed directly on raw sense-data by the mind' (p. 96), and although the constructions might change over time, the data themselves do not. The data can therefore be captured in a 'pure' or 'neutral' observation-language, and will be equally available to all concerned.

Against this, Kuhn makes several moves. First, he insists that recent research in philosophy, psychology, linguistics and art history suggests that this traditional philosophical paradigm must be flawed. The psychological literature on perception, in particular, suggests that 'something like a paradigm is prerequisite to perception itself',[21] and 'the history of science would make better and more coherent sense' if one could suppose that scientists occasionally experience the sort of Gestalt-shifts in perception which Hanson, for example, drew attention to (p. 113). Kuhn recognizes the limitations of this first move, though, admitting that the psychological experiments in question can't be any more than suggestive. Perhaps simply because they don't directly concern scientific observation, they can't demonstrate that *this* kind of observation has the characteristics of perception displayed in the experiments. His second move, therefore, is to suggest that these psychological experiments must be supplemented by *historical* example to make them relevant. And sure enough, according to Kuhn, the history of science also suggests that the Western epistemological paradigm is flawed.

This doesn't mean that *all* the characteristics of Gestalt-psychological experiments are reflected in scientific observation. Kuhn is clear that they aren't, and explicitly notes several features of the latter that don't correspond to the former (p. 114). Even if perceptual switches do accompany paradigm-changes, for example, scientists themselves will not be directly aware of them and will be unable to testify to them. What sorts of transformations in the scientist's research-world *can* the historian who believes in such changes discover, then? Kuhn proceeds to give examples from the history of astronomy, electricity, chemistry and mechanics. In each of these cases, he describes the facts or realities in question as themselves having changed as a result of theory-change. So, for example, when Lexell's suggestion that the orbit of Uranus was planetary was accepted, Kuhn says, 'there were several fewer stars and one more planet in the world of the professional astronomer' (p. 115). Lavoisier 'saw oxygen where Priestley had seen dephlogisticated air' (p. 118). And until the scholastic impetus theory was invented, 'there were no pendulums, but only swinging stones, for the scientist to see. Pendulums were brought into existence by something very like a paradigm-induced Gestalt switch' (p. 120).

Unsurprisingly, these descriptions have been challenged on the very reasonable ground that in each case what was involved was a shift of attention or a change of categorization *within* a single perceptual world, rather than a change to a *new* world. One might, for example, describe matters by saying that the earlier astronomers *saw* a planet but *interpreted* it as a star, and that Aristotle saw a pendulum but interpreted it as a stone undergoing constrained fall. Alternatively, one might say that what earlier astronomers saw and categorized *as* a star, later astronomers saw and categorized *as* a planet. Or that what Aristotle saw *as* a stone undergoing constrained fall, Galileo saw *as* a pendulum. Kuhn, on this view, confuses determining what observations should be made with determining what the result of the observations will be. (And it's notable that Kuhn himself does sometimes describe things in this more accommodating way, e.g. where he says that different paradigms made different data *accessible to* different scientists (p. 123).)

Against the specific suggestion that when scientists change their minds about what they observe, they do so merely by *reinterpreting* stable observational data, Kuhn insists that the data are not stable (p. 121), and that the *process* by which the individual or community makes the transition isn't one that resembles interpretation (pp. 121–2). Although the first point is rightly regarded as contentious, the second is more promising. Interpretation is a rule-governed process which proceeds from agreed data and finds the best construction to put on them, 'a *deliberative* process by which we *choose* among alternatives', as Kuhn says in his Postscript (p. 194, emphases added). But the scientists in question, Kuhn points out, were not in a position to first agree on some more basic description of data and then proceed to interpret them. Interpretation was unnecessary for them, and no such process took place. A new paradigm is born in what Kuhn calls a 'flash of intuition' (p. 123), not a process of interpretation, since interpretation is too *piecemeal* and *intellectualized* a process compared to the *holistic* and *automatic* phenomenon Kuhn has in mind. Kuhn doesn't deny that scientists do engage in interpretation, or that reasoned reflection is involved in the development of a paradigm. He recognizes that scientists interpret observations and data, for example (p. 122) (and later he suggests that they interpret nature itself (p. 144)). These interpretations, though, like the more nearly successful attempts to produce a pure or neutral observation-language (pp. 125–6), but unlike those conceived under the usual epistemological picture, *presuppose* a paradigm. So they don't provide the usual epistemological story with what it requires.

At one point Kuhn suggests that rather than being an interpreter, 'the scientist who embraces a new paradigm is like the man wearing inverting lenses. Confronting the same constellation of objects as before and knowing that he does so, he nevertheless finds them transformed through and through in many of their details' (p. 122). Here, though, Kuhn hasn't quite got his story straight, since this description conflicts with what he says later, where the inverting-lenses subject is characterized as having *different* retinal impressions and yet seeing the *same* thing as before (p. 127). His better suggestion, more relevant to paradigm-change, remains the idea that the phenomenon of *Gestalt-switches* shows that two people with the *same* retinal impressions can see *different* things (ibid.).

Kuhn credited his interest in Gestalt psychology to his 1947 Aristotle epiphany (Sigurdsson 1990, p. 20). But although, like Wittgenstein, he had read some of the Gestalt psychological literature on perception (p. vi, ET, p. xiii), it's difficult to see what more he took from it than the basic idea of Gestalt-switches. (Further, although he thought of himself as synthesizing recent developments in various fields, the heyday of Gestalt psychology was long over by the time Kuhn was writing SSR.) He seems to have been more influenced by the *Philosophical Investigations*,[22] Part II of which contains an extended (albeit unfinished) discussion of what Wittgenstein calls 'noticing an aspect'. That discussion starts with Wittgenstein distinguishing between two senses of the term 'see':

The one: 'What do you see there?' – 'I see *this*' (and then a description, a drawing, a copy). The other: 'I see a likeness between these two faces' – let the man I tell this to be seeing the faces as clearly as I do myself. (Wittgenstein 1958, p. 193)

Wittgenstein suggests there's a difference of *category* between the two 'objects' of sight here, because it's perfectly possible for the latter person to 'notice in the drawing the likeness which the former did not see' (ibid.). So what is seen in the first case (a public object of a representational nature) isn't the same *kind* of thing as what is seen in the second (a resemblance, similarity, or aspect). Equivocation on 'what is seen' can produce the *appearance* of a paradox (what is seen has changed, but what is seen has *not* changed). This appearance dissolves, though, when one separates the two senses, as Wittgenstein does.

Seeing the object differently, even though it hasn't changed, is what Wittgenstein calls 'noticing an aspect'. This phenomenon goes with the question '*How* do you see it?' (Whereas the question '*What* do you see?' can be understood as asking for *either* the first kind of 'object' of sight *or* the second.) As a result, Wittgenstein associates the second sense of 'see' with the idea of '*seeing as*', and moves freely between the two. And he clearly considers this second sense of 'see' closer to *thinking* and to *interpretation* than the first sense is.

For Wittgenstein, ordinary perception *isn't* aspect-perception, and aspect-perception is *occasional*, not ubiquitous. He denies that it makes sense to say, at the sight of objects as ordinary and familiar as a knife and fork, for example, that one is seeing those *as* a knife and fork (ibid., p. 195). Aspect-perception is bound up with *change of* aspect, which involves *noticing a resemblance*, so it can't be *ubiquitous*. Aspect-perception also goes with *trying*, i.e. it is subject to the will. But no such act of will is involved in seeing objects such as a knife and fork.

Wittgenstein stresses the distinction between having a continuous dispositional attitude (taking something to be an X) and the sudden and occasional 'lighting up' of an aspect or noticing of an aspect (ibid., p. 194). In the latter cases, when one notices a resemblance, a change is wrought in one's perceptual experience. Wittgenstein specifically denies that 'seeing', in this second sense, is a matter of picking up information (ibid., p. 197).[23] Of course, that *is* what we're usually concerned with in the scientific case. But Kuhn, I think, is trying to advert to an aspect of scientific observation that the picking-up of information *presupposes*.

Let's look at SSR's discussions of perceptual switches in the light of Wittgenstein's comments. Back in section VI Kuhn insisted, of Bruner and Postman's experiment, that 'one would not even like to say that the subjects had seen something different from what they identified' (p. 63). He doesn't say why, but his claim is only plausible if one is using Wittgenstein's *second* sense of 'see'. If someone sincerely says 'I see a likeness between these two faces' no-one else can really *correct* this. It isn't empirically falsifiable. This is because, according to Wittgenstein, such an utterance is an *avowal*, a spontaneous and natural reaction to what's seen, rather than a *description* of it. The question then is: in the case of utterances produced in the course of scientific observation, should scientists be thought of as *avowing* what they perceive, rather than describing it? Are scientific observation-statements to be thought of as thus unimpeachable? Perhaps Kuhn is suggesting so, in the limited sense that *these* scientists (like members of a 'primitive tribe') might be incapable of being brought to recognize that what they 'saw' wasn't really there. In particular, their basic intuitions as to what features of a situation are *similar* are not empirically falsifiable, and thus not plainly empirical.[24] In so far as this is the case, their *way of doing science*, their *natural reactions*, may be fixed. As a consequence, they cannot (at that time, at least) undergo the Gestalt-switch in question.

Even if one goes this far with Kuhn, though, some important qualifications must be entered. First, it's important to his own case that not *all* scientists are in this position. There have to be *some* (those whose work turns out to have been revolutionary) whose perceptions *can* be changed (indeed, it may be important, *pace* Kuhn, that their perceptions are *reversible*, so that they can move in thought from one paradigm to another, and back again). This creates a problem for any way of describing the situation that relies *exclusively* upon the second sense of 'see'. Second, the availability of the first sense of 'see' means one *can* still legitimately describe the uncorrectable scientists in question as having been in error, as not really having seen what they *thought* they saw. Kuhn correctly notices the second sense of 'see', to which Wittgenstein drew attention, and in which (roughly) one sees all and only what one thinks one sees.[25] But Kuhn seems sometimes to forget that there's also *another* sense, the more ordinary sense, in which one can *only* see what is there to be seen, the verb 'to see' being a *success*-verb. There is, *pace* Kuhn, nothing wrong with insisting that Bruner and Postman's subjects only *thought they saw* (e.g.) the red four of spades, *just because* no red four of spades was there to be seen. (Indeed, Bruner and Postman themselves were careful to talk about what their subjects saw or thought they saw (e.g. p. 210).)

In his judgement about the Bruner and Postman experiments, Kuhn also takes advantage of this *factive* aspect of 'see', its being a success-verb, in order to imply that what is seen must be present. But, when talking of scientists whose paradigm or conceptual scheme we don't share, enforcing this implication (by insisting that this is what they *saw*) may be inappropriate, by committing *us* to the existence of things we don't believe in. In such cases, we sometimes need to be able to say that scientists *thought* they saw things which (by our lights) weren't really there.

One can, of course, say that the subjects in question saw the incongruous test-card (e.g. a black four of hearts) *as* the four of spades. But this is *seeing as*, not seeing. The former doesn't (or needn't) exhibit the factive feature of the latter (that if someone can truly be said to see X, X must actually be visually present to them). Kuhn specifically makes it clear that he *isn't* relying on a thesis about 'seeing as'. He notes that some (e.g. Hanson) who have noticed the feature of paradigm-transition he is interested in (its being a reconstruction of the field from new fundamentals), have likened it to a change in Gestalt: 'the marks on paper that were first seen as a bird [being] now seen as an antelope, or vice versa' (p. 85). *But he finds this parallel misleading*, since scientists 'do not see something *as* something else; instead, they simply see it' (ibid.). Describing scientists as seeing something *as* oxygen, Kuhn seems to feel, doesn't do justice to the unhesitating nature of their perception, or to the way they would avow it. As Wittgenstein says, 'I should not have answered the question "What do you see here?" by saying: "Now I am seeing it as a picture-rabbit". I should simply have described my perception: just as if I had said "I see a red circle over there"' (Wittgenstein 1958, pp. 194–5).

This, though, means that what Kuhn is interested in isn't quite what Wittgenstein meant by 'aspect-perception'. Wittgenstein insists that aspect-seeing is *sudden*, *occasional* and *reversible*. Kuhn agrees that the conversion experience he is concerned with is 'relatively sudden' (pp. 122, 150), albeit more gradual than a Gestalt-switch (p. 111), but treats it as *ubiquitous* (for the scientists concerned) and explicitly says it is *irreversible* (pp. 85, 111). Nevertheless, because he only ever claims that the Gestalt-switch is 'a useful elementary prototype' of what happens in 'full-scale' paradigm-shift (pp. 85, 111, i.e. shift in disciplinary matrix), Kuhn isn't committed to there being anything more than an *analogy* between the two kinds of case. The question is: is there *enough* of an analogy?

What Kuhn is talking about is more like the continuous dispositional attitude of simply *taking something to be an X* which Wittgenstein *distinguishes* from aspect-perception. Kuhn would presumably want to go further than this (as he does in his move against Hanson) and say that scientists don't merely 'take [something] to be oxygen' (one can hear him asking: what would this 'something' be, anyway?): they *see* oxygen. This would only be true if one could use the *second* sense of 'see' in this connection. But when the verb 'to see' takes an object like 'oxygen' (rather than a more nebulous 'object' like 'a resemblance', 'a similarity', or 'an aspect') only its *first* sense really makes sense. *That* sense of the verb, though, being factive, implies the existence of a perceived phenomenon which comes (as a result of the scientific revolution in question) to be publicly checkable, meaning that any claim to see something of *that* kind is potentially corrigible, and we cannot simply grant all claims to the effect that such phenomena are seen. Here it seems that Kuhn has *conflated* aspects of the two senses of 'see' that Wittgenstein drew attention to: if one takes advantage of the factive aspect of 'see', one cannot at the same time rule out the possibility that what the subject sees isn't what they claim to see.

In summary, Kuhn may have thought he was following Wittgenstein, but his treatment agrees with Wittgenstein's only in part. The most important point of agreement is that when a new aspect of a seen object is noticed, a change is wrought in one's perceptual experience, and this is not to be thought of as the acquisition of some new item of information. A second point of agreement is that, in the second sense of 'see', people should be thought of as *avowing*, rather than describing, what they see. Scientific observation-statements may therefore be unimpeachable in the limited sense that there may be no purely argumentative way to bring their producers to recognize that what they 'see' isn't really there. Likewise, the basic intuitions of similarity which partly constitute a paradigm are not open to straightforward empirical falsification. Like a visual illusion, the fact that one thing strikes one as similar to another can persist through any demonstration that those things are not really related in that way. Thirdly, Kuhn agrees with Wittgenstein that perceivers don't naturally avow their perceptions using locutions such as 'seeing as' and 'taking as'. They just describe what they perceive. Using such locutions suggests (for Wittgenstein) that one's perceptions have changed, or (for Kuhn) that there is some hesitation on the perceiver's part, and these implications are, in the cases concerned, incorrect.

But the *differences* between Wittgenstein and Kuhn are as important as the similarities. First, although Kuhn concentrates on some of the same uses of 'see' as Wittgenstein does, he doesn't explicitly say that they form a separate sense of the word. In fact, the only time he addresses the *first* sense of the word, the one Wittgenstein treats as its ordinary sense, Kuhn suggests that it is 'questionable' (p. 55), but without really following this up.

Second, Kuhn specifically *dissociates* the uses of 'see' he is interested in from the phenomenon of 'seeing *as*', with which Wittgenstein connects them. Where Wittgenstein takes 'seeing as' to be part of his topic, Kuhn denies that it's right to describe scientists' perceptual experiences with this phrase.[26]

Third, where Wittgenstein *distinguishes* between the sudden 'lighting up' or noticing of an aspect, and the continuous dispositional attitude of taking something to be an X, Kuhn assimilates them.

Fourth, where Wittgenstein is very clear that 'noticing an aspect' is something that happens only occasionally, Kuhn takes the kind of seeing that he's concerned with to be ubiquitous (for the group of scientists in question). Relatedly, where Wittgenstein takes noticing an aspect to be subject to the will, and treats 'seeing as' as a neighbour of *interpretation*, Kuhn takes the kind of aspect-perception he's concerned with to be entirely *natural* for the scientists involved, and in no way under their voluntary control. For him, it is to be contrasted with interpretation.

Fifth, as we have seen, Wittgenstein's concentration on the classic Gestalt-figures leads him to consider changes of aspect as *sudden*, *holistic* and *reversible*. Kuhn claims that the scientific transitions he has in mind are sudden, albeit 'more gradual' than Gestalt-switches, and denies that they are reversible, but retains the idea that they are *holistic*.

None of this need mean that Kuhn is wrong. It may be, to reiter-
ate, that he's simply talking about a somewhat different phenomenon
from Wittgenstein, since he's clear that he's using the idea of Gestalt-
switches merely as a metaphor or 'prototype'. But Kuhn would have
to show, first, that there *is* a sense of 'see' answering to his description. This seems unlikely. There just doesn't seem to be a *single* sense
of this verb which will give Kuhn everything he wants, i.e. both the
incorrigibility of judgements like 'I see X' *and* the factive implication
that if one sees X, X must really be present. More importantly, even
if there were such a sense, the availability of the *first* sense of 'see'
means Kuhn cannot continue to claim that his opponent is *wrong* to
insist that experimental subjects or past scientists didn't really see
what they thought they saw. As Wittgenstein recognized, even if I
would not have described *myself* as seeing a figure *as* a picture-
rabbit, nevertheless 'someone else could have said of me: "He is
seeing the figure as a picture-rabbit"' (Wittgenstein 1958, p. 195). It
may be no part of the *historian's* brief to make such judgements, but
that doesn't mean they cannot legitimately be made.

The Gestalt-switch is only supposed to be a metaphor, softening
us up for a conclusion, and Kuhn could dispense with it (as indeed
he did in his later work). But that conclusion, the idea of 'world-
changes', which is the other central idea of section X, is just as problematic and less dispensable.

During scientific revolutions, Kuhn says, scientists 'see new and
different things when looking with familiar instruments in places
they have looked before' (p. 111). 'Insofar as their only recourse to
that world is through what they see and do', they respond to a
different world (ibid.). And historians of science (by which Kuhn
means those in the wake of the historiographic revolution) are
tempted to say that the world *itself* changes. Later on, what the historian is initially 'tempted' to say becomes *mandatory*: we are told
that we *must*, despite the difficulties, learn to make sense of statements like 'Scientists worked in a different world' (p. 121).

How does Kuhn understand the term 'world'? Paradigm-changes,
he tells us, cause scientists to see '*the world of their research-
engagement*' differently (p. 111, emphasis added). The student of
science comes to inhabit '*the scientist's world*' (ibid., emphasis
added), which means seeing what s/he sees and responding as s/he
does. *This* world is not 'fixed once and for all by the nature of the
environment, on the one hand, and of science [in general], on the

other' (p. 112). It is determined *jointly* by the environment and the *particular* normal-scientific tradition. And it is '*the world of [the scientist's] research*' that seems incommensurable with the one she or he inhabited before (ibid., emphasis added). When the suggestion that the orbit of Uranus was planetary was accepted, for example, 'there were several fewer stars and one more planet in *the world of the professional astronomer*' (p. 115, emphasis added).

These phrases indicate that Kuhn didn't take the term 'world' with the ontological seriousness that many contemporary philosophers want to imbue it with. After all, if the student of science comes to inhabit '*the scientist's world*', those who *aren't* scientists don't even live in the same world as scientists! By 'world', then, Kuhn means '*world of the scientist's research-engagement*' (just as one might talk of the pig-breeder's world).[27] Perhaps he did not initially realize how weighty an issue concepts such as *a* world and *the* world raise for philosophers. What they mean by 'the world' Kuhn usually calls 'the environment' or 'nature'.[28]

Kuhn, I suggest, vacillated between two kinds of view. On the one hand, there's what one might call a modified Kantian picture. Immanuel Kant was one of the few philosophers Kuhn had read while a physics undergraduate, and had an enormous impact on him (RSS, p. 264). But, like any good post-Kantian, Kuhn couldn't accept the static nature of Kant's 'categories'. According to his modified Kantian picture, then, successive paradigms function like conceptual schemes ('conceptual spectacles' as Kuhn says (ibid., p. 221)), through which the world is conceptualized (that is, experienced, perceived, thought about) by different groups of scientists. In his later years, Kuhn explicitly thought of his views this way, calling them a Kantianism 'with moveable categories' (ibid., p. 264), or a 'post-Darwinian Kantianism' (ibid., p. 104). On this view, there *is* a real world and, of course, it doesn't change as a result of our reconceptualizing it. What changes are the *research worlds* of the scientists involved. This reading of Kuhn has been most thoroughly elaborated and defended by Paul Hoyningen-Huene, who calls the worlds that change 'phenomenal worlds' (Hoyningen-Huene 1993, part II). These are constituted by the conceptual apparatus of a paradigm, notably its similarity relations, and change when that paradigm changes. This reading of Kuhn's work certainly has the virtue of meaning that it is not necessary to write off his heavily qualified suggestions that there is 'no one world' in which scientists work. But

KUHN'S *THE STRUCTURE OF SCIENTIFIC REVOLUTIONS*

the main problem with it, of course, is that it simultaneously makes *the* world, the unchanging one, into an ineffable *Ding-an-sich* (thing-in-itself), a something about which nothing can be known. Thus it still retains what most post-Kantian philosophers, as a matter of fact, came to reject. A further worry is that it seems rather too close to the usual idea that successive scientific communities develop different *interpretations* of something stable (the world-in-itself), an idea Kuhn clearly wanted to get away from when the stable items in question were data.

Kuhn struggled with this Kantian view throughout his career, at some times repudiating it, then taking it up again, but then later on again rejecting it. It certainly does do justice to most of what Kuhn says. But that problem about the ineffability of the world as *Ding-an-sich* remains, and it was sometimes enough to put Kuhn himself off. When he repudiated modified Kantianism, it was precisely because it made the world 'in principle unknowable' (RSS, p. 207). Even within this section of SSR, his reference to the world as a '*hypothetical* fixed nature' (p. 118, emphasis added) surely expresses some scepticism about the notion. In a spirit of charity, therefore, because of these problems, other possible readings are worth exploring.

One interpretation well worth considering is Ian Hacking's, according to which Kuhn doesn't deny the mind-independent existence of individual things, but he does deny that the ways we *classify* scientific things into categories reflect facts about the way the world is independently of us. There are real individuals (the world of individuals doesn't change), but no real theoretical *kinds*. What happens when the 'world' changes, on this view, is that one world of theoretical kinds, one way in which we systematically divide up (some part of) the world, is replaced by another. Hacking calls this view 'revolutionary transcendental nominalism' (see, e.g., Hacking 1979, and his essay in Horwich 1993). It's 'revolutionary' because the kinds can *change*, 'transcendental' because it concerns only *theoretical* kinds, and 'nominalism' because it denies that *such* kinds are real, while allowing that individuals are.

Kuhn, though, responded to Hacking by denying that individual things have mind-independent existence, insisting that what was needed was 'a notion of "kinds" . . . that will populate the world as well as divide up a preexisting population' (Kuhn in Horwich 1993, p. 316). That response does resuscitate the worry about idealism.

READING *THE STRUCTURE OF SCIENTIFIC REVOLUTIONS*

There may also be a problem pertaining to Hacking's talk of empirical and 'trans-empirical' kinds. There is nothing in Kuhn to suggest that he would not have recognized a rough spectrum of this sort on which kinds fall. But the view Hacking outlines seems to require something more problematic, a single clear-cut *distinction* between empirical and 'trans-empirical' kinds. It's difficult to see how there could be a single such fixed distinction, or how it could be motivated from Kuhn's perspective, given how unenthusiastic he was about similar distinctions when they were relied on by his logical empiricist opponents.

Kuhn's occasional but inconstant *anti*-Kantian intuitions suggest looking further afield to show what he might have said without being saddled with the sort of problems Hoyningen-Huene's and Hacking's readings raise. There is a more thoroughgoing picture that Kuhn is tempted by (in his anti-Kantian moments), but doesn't quite commit to. A powerful expression of it appears in Peter Winch's influential book *The Idea of a Social Science* (which Kuhn may well have known, although I am not suggesting any direct influence):

> Our idea of what belongs to the realm of reality is given for us in the language that we use. The concepts we have settle for us the form of the experience we have of the world. It may be worth reminding ourselves of the truism that when we speak of the world we are speaking of what we in fact mean by the expression 'the world': there is no way of getting outside the concepts in terms of which we think of the world. . . . The world *is* for us what is presented through those concepts. That is not to say that our concepts may not change; but when they do, that means that our concept of the world has changed too. (Winch 1958, p. 15)[29]

If this is right, Kuhn's 'world-changes' thesis is not quite the right way to put things. It's *not* that the world changed as a result of its reconceptualization. After all, since the only thing *we* can mean by 'the world' is given by our best (i.e. current scientific) account of its nature, the world just isn't the *sort* of thing that could change as a result of conceptual change. The world of stars, planets, chemical elements, organisms, etc. *as we now conceive those things* is as solid a world as any 'realist' could *ever* want! It isn't merely 'phenomenal', but it is, of course, the *only* world. Kuhn, I suggest, is tempted to think that the world is what is meant by 'the world'. But it's not. The right way to put this view is to say (as Winch does) that the world is what *we* mean by 'the world'. Although his 'world-changes' thesis is obscure, perhaps what Kuhn *meant* by it was that as a result of paradigm-change, what 'the world' meant changed. The ancients, that is, meant something different by 'the world'; they had a different *concept* of the world. This *doesn't* really mean that the world changed, or that it's 'culture-dependent'. Science *can* still be seen as telling us more and more about a mind-independent world.

Of course, this sort of view, which has been developed by Robert Arrington under the name 'conceptual relativism',[30] has one implication that Kuhn *isn't* going to want, and which may be the reason he doesn't consistently commit to it. It implies that the ancients weren't fully epistemically in touch with the world. (Just as they had *a* morality, but not *morality*, not what we mean by 'morality', they had *a* world, but not *the* world. One has to be *careful* in saying these things, though.) This is perfectly compatible with agreeing with Kuhn that although past scientists were working within a conceptual scheme which we can't help but regard (rightly) as inadequate, they weren't simply *mistaken* or in error. A mistake or error is something committed by a person making a claim using the resources of a particular conceptual scheme, whereas what's in question here is the use of *those* resources in the first place.[31] Neither did past scientists behave irrationally: they were simply *different*, and their paradigm was flawed. Nevertheless, the perspective I have in mind doesn't license the generally-perceived implication of 'tolerance' for the views of the past scientists in question. Our world, and nothing else, is *the* world.

What one should say is perhaps this. Past scientists weren't, as it were, in touch with *nothing*. They were in touch with what they meant by 'the world'. (Perhaps, then, one might say they were in touch with *their* world.) But they used conceptual schemes that threw up various insoluble problems, and didn't allow the development of true general statements. Thus, since we aim, in science, for theories which will allow us to do that, we no longer want to use those schemes. Using those conceptual schemes, though, the ancients did manage to say plenty of true things, things that were true of the objects in question, even though those objects do not add up to what we mean by 'the world'.

This perspective fits, I think, with MacIntyre's attempt to rescue Kuhn from the suspicion that the epistemological crises he adverts to (the larger scientific revolutions) would involve changes in our narrative of history so total as to be irrational:

> What the scientific genius, such as Galileo, achieves in his transition . . . is not only a new way of understanding nature, but also and inseparably a new way of understanding the old science's way of understanding nature. It is because only from the standpoint of the new science can the inadequacy of the old science be characterized that the new science is taken to be more adequate than the old. It is from the standpoint of the new science that the continuities of narrative history are re-established. [. . .] I am suggesting, then, that the best account that can be given of why some scientific theories are superior to others presupposes the possibility of constructing an intelligible dramatic narrative which can claim historical truth and in which such theories are the subject of successive episodes. It is because and only because we can construct better and worse histories of this kind, histories which can be rationally compared with each other, that we can compare theories rationally too. Physics presupposes history and history of a kind that invokes just those concepts of tradition, intelligibility, and epistemological crisis for which I have argued. (MacIntyre 1977, pp. 467, 470).

The final thing to be said about this alternative picture, though, is that it's no longer Kantian, but perhaps more like the thoroughgoing view of Kant's successor, G. W. F. Hegel, and some of his followers.[32]

If this *is* a kind of relativism, it's a relativism about *concepts*, not about truth or truths. If Kuhn really did mean to go further than this, as many commentators think, to be a relativist about *truth*, I think we ought not to follow him. The temptation to say that the claims of past scientists were true *relative to something* (*their* paradigm or conceptual scheme) is, I think, indefensible. Truth just isn't relative, since, for any proposition 'p', p's being true is simply a matter of p being the case. And what is the case about natural-scientific phenomena (as opposed to what is *thought* to be the case) can't differ from person to person, paradigm to paradigm, or conceptual scheme to conceptual scheme. That's just what it means to say that *the* world doesn't change merely in response to how it is thought of. Relativism *is* usually taken to be such a view, and most objections to it derive from this. But it need not be. The pluralism about conceptual schemes envisaged here, not being a relativism about truth at all, avoids the usual objections on the score of self-refutation, for example.

Arrington (1989, p. 260) takes this 'conceptual relativism' to mean that scientific concepts cannot be justified metaphysically, pragmatically or instrumentally. Here I think one might legitimately disagree. Kuhn's own perspective, I shall suggest in section 8, licenses the idea that even though paradigms and their constituent concepts cannot have a *metaphysical* justification, they may nevertheless have a *pragmatic* or instrumental one. Pluralism or relativism about concepts is perfectly compatible with the idea that science makes progress (and even with the idea that scientific claims are true-or-false).

What Kuhn says about world-changes in his Postscript to SSR again suggests that the 'worlds' he is concerned with are indeed 'phenomenal worlds'. He contrasts the public stimuli people receive in perception with the 'sensations' they have as a result, and argues that it is groups of people who receive the same stimuli but have different 'sensations' who live in different 'worlds' (pp. 192–3). (Hoyningen-Huene 2003, chapter 2, raises serious problems with this 'stimulus ontology'.)

This section of Kuhn's Postscript also clarifies his distinction between interpretation and perception, sharpening his critique of the usual epistemological 'paradigm'. Interpretation, he says, and as we have already noted, is 'a deliberative process by which we choose among alternatives as we do not in perception itself' (p. 194). In interpretation we deploy criteria and rules. But the processes and techniques by which scientists-to-be are trained to see the same things when confronted with the same stimuli don't involve learning *rules*, since their recognition of those things isn't voluntary. Only *after* one has perceived the thing in question can one engage in the rule-governed process of interpretation or analysis. Interpretation, as Kuhn puts it, 'begins where perception ends' (p. 198).

Study questions

1. Is the Gestalt-switch metaphor more appropriate to science education than to paradigm-change? Why can historians of science, but not scientists, attest to the perceptual changes Kuhn has in mind? Would the phenomenon of inverting-lenses be a better analogy to scientific perceptual changes than that of Gestalt-switches?

2. What different senses of verbs like 'to see' and 'to perceive' are there? Is there a single sense of such verbs answering to Kuhn's needs?
3. Do scientists avow what they perceive, rather than describe it? If so, does this mean that one must accept their avowals? Is there a sense in which (correctly produced) scientific observation statements are unimpeachable? Should Kuhn have been concerned with seeing-as, rather than with seeing?
4. Which kind of 'worlds', if any, change during scientific revolutions: phenomenal worlds, worlds of kinds, the world in itself, or some other kind of worlds? Or is it merely what we mean by 'the world' that changes? If the world is unchanging, and therefore so are 'stimuli', why should we think that *data* can change? Is the world unknowable?
5. What, if anything, was Kuhn a relativist about?
6. Would a wider conception of interpretation succeed in legitimately reinstating the usual story according to which scientists interpret the same data, but do so in different ways? Should interpretation be contrasted with perception?

SECTION 7: THE INVISIBILITY AND RESOLUTION OF REVOLUTIONS

In order to finally convince readers of the existence and nature of scientific revolutions, section XI investigates why they have not generally been recognized. It traces this to the nature of scientific textbooks, but then explains why this invisibility of revolutions is a legitimate functional feature of such texts. Section XII addresses the question 'How do scientific revolutions close?' Because Kuhn denies that confirmation or falsification play anything like the role philosophers of science usually think they do, he has to explain what takes their place. Paradigms are never tested (and thus never confirmed or falsified) in isolation, but only once a crisis has set in, and only as part of the process of paradigm-competition.

Kuhn has already conceded that scientific revolutions do seem revolutionary at least 'to those whose paradigms are affected by them' (p. 93, cf. p. 50). But he now takes it to be part of his task to explain not just why scientific revolutions occur, but also why they have been generally thought *not* to occur; why they have been 'invisible'. His answer is that:

Both scientists and laymen take much of their image of creative scientific activity from an authoritative source that *systematically disguises* – partly for important functional reasons – the existence and significance of scientific revolutions. (p. 136, emphasis added)

The source in question is the scientific textbook, together with popularizations and the philosophical works based on such texts. Because their task is not to describe the upheavals which constitute scientific revolutions, but to describe the *results* of such revolutions, the existing scientific wisdom on a given topic, such texts seek to convey a finished body of normal-scientific work, a *fait accompli*. (Kuhn gives a useful characterization of the *kinds* of books sciences do and don't use in ET, pp. 228–9.)

Because science texts have to be rewritten in the aftermath of each scientific revolution in that field, they will, in so far as they portray the history of their own subject, 'inevitably disguise not only the role but the very existence of the revolutions that produced them' (p. 137). In fact, the historiography embodied in such texts is almost inevitably what Kuhn and others called *whig historiography*. This is a popular way of seeing the past taken by certain historians working before the historiographic revolution (principally those writing the history of political institutions). When applied to the history of science, it involves rewriting that history in order to make the development of scientific activity look not merely *progressive* but *cumulative*. Whig historians study the past primarily with reference to the present. They not only see the history of science as the history of progress towards our current scientific world-picture, but they also try to divide up the past history of science into those aspects which prefigure current scientific views, and those which do not. They treat the former as the more significant, as anticipations of what is now known, and the latter as unfortunate aberrations. By favouring and focusing primarily on the former, the result is that the written history of science comes to be a story of how contemporary science builds on and smoothly expands the insights of past scientists.

While Kuhn wasn't exactly surrounded by whig historians of science (he later remarked that there were perhaps fewer than six professional historians of science in the US at the time!), he did count Harvard's main historian of science, the Belgian George Sarton (1884–1956), as such (see RSS, pp. 275, 281–2). In his 1936 book *The Study of the History of Science*, for example, Sarton repeats some ideas which, he says, he had published in various forms since 1913:

Definition. Science is systematized positive knowledge, or what has been taken as such at different ages and in different places.
Theorem. The acquisition and systematization of positive knowledge are the only human activities which are truly cumulative and progressive.
Corollary. The history of science is the only history which can illustrate the progress of mankind. In fact, progress has no definite and unquestionable meaning in other fields than the field of science. (Sarton 1936, p. 5)

Sarton was by no means as one-dimensional a figure as this quote, and Kuhn's very occasional mentions of him, make him appear. Nevertheless, Kuhn was convinced that there was 'a sort of history of science to do that Sarton wasn't doing' (RSS, p. 282).

In steadfastly opposing what he thought of as whig historiography, Kuhn faithfully followed his *maître*, Koyré, and Butterfield, who had engaged in a famous attack on whiggism (Butterfield 1931), and who was himself already greatly influenced by Koyré. But their critique was rather blunt and indiscriminate. Whig historiography certainly errs in trying to understand past views in terms of present ones, and in focusing exclusively on those aspects of past theories which appear to anticipate present theories. Its tendency to divide past scientists into heroes and villains along the same lines is equally problematic. And it certainly involves the risk of portraying as anticipations what are in reality merely misleading analogies. Koyré and Kuhn were quite right to insist that the historian should try to 'get inside the heads' of past scientists, to see past scientists as people of their *own* time, rather than ours, and to try to interpret past science as having its own coherence and integrity. But Butterfield's insistence that the historian's main task is to elucidate the *unlikenesses* between past and present, to destroy any analogies we may have thought to exist, seems an over-reaction to whiggism. And whig historiography does not err, surely, in trying to evaluate the merits of past theories, or even in suggesting that the history of science is a story of progress. Kuhn himself, as we shall see, agreed with judgements of this kind.

Whig historians picture older scientists as having worked on the same set of problems that current science is concerned with, in accordance with the same set of principles. No wonder, then, that those not directly engaged in science, taking their image of science mainly from science texts, conceive of the development of science as a process of accumulation. The crudest and most obvious way in which to interpret this phenomenon would be as an illustration of the familiar political maxim: 'the winners write the history books'. Some sociologists have understood Kuhn to be making this complaint. But Kuhn argues to the contrary that this phenomenon should not surprise us, or be reckoned a disgrace upon the objectivity of science itself. He insists that there is an excellent *functional* reason why any historical content of scientific textbooks, in particular, should be presented in this way. Sciences, as they develop, rely increasingly on textbooks. Because the function of such textbooks is to educate aspiring scientists with *current* paradigms, scientific education would simply be less effective if textbooks weren't partial and even misleading in this respect. The central reason why they must ignore, forget, or at least re-interpret the works of their heroes is that textbooks exist to teach people *science*, not about the *history* of science. Science aims to teach us how to predict, understand and manipulate things *now*. But since our ways of dealing with things have improved in terms of their instrumental success, how we dealt with them in the past is of no interest to the education of *new* scientists unless it is *still* the best way we have.

Textbooks like this, then, Kuhn argues, are one indicator of what sets the developmental pattern of mature science apart from the development of any other field of creative activity. No other kind of intellectual field has texts of quite this kind. In other words, it's because science is genuinely *progressive* that good pedagogical practice goes hand in hand with bad historiography.

Over against this historiographically challenged image, Kuhn sets his own story, which portrays *changes*, rather than continuities, as the key components in the development of science. If what he has said so far is correct, we already know that scientific revolutions, although they bring in train new empirical discoveries, cannot themselves consist merely in such discoveries. The *puzzles, problems, facts* and *theories* of contemporary normal science are not eternally existing, but only really come into being after a scientific revolution. During such a revolution, the *questions* scientists ask, together with the kinds of *answers* they deem acceptable, are subtly reformulated. One cannot, however, *tell* this by looking merely at what scientists say. What they say (such as the way Boyle characterized the notion of an 'element') may remain constant across a scientific revolution, but this doesn't matter. Such definitions have little scientific content, being mere 'pedagogic aids'. 'The scientific concepts to which they point gain full significance only when related, within a text or other systematic presentation, to other scientific concepts, to manipulative procedures, and to paradigm applications' (p. 142). Concepts like that of an element, then, can't be invented *independently* of context. But, *given* the context, they rarely require invention, since they are already at hand.

These last few pages of section XI certainly show Kuhn flirting with the idea that various notions are *conception-dependent*. No-one, I think, should quibble with the idea that *theories* are the results of human conception. The concepts of puzzle and problem, too, have enough reference to conception built into them in order for us to find unproblematic the idea that they come into and go out of existence as a result of conceptual or intellectual shifts. Things are otherwise, though, with the notion of a fact. What textbooks present as theories, we are told, 'do "fit the facts", but only by transforming previously accessible information into facts that, for the preceding paradigm, had not existed' (p. 141). Saying this may not involve treating facts as created, because of that qualifier, *'for the preceding paradigm'*. Kuhn may just mean that scientists within that preceding paradigm failed to recognize such facts. But he goes on to say that theories 'do not evolve piecemeal to fit facts that were there all the time. Rather, they emerge together with the facts they fit from a revolutionary reformulation of the preceding scientific tradition' (ibid.). To take *this* seriously means thinking of facts as the kind of thing which 'exist' as a result of being created *by* scientific revolutions.[33] A minor problem with this is that facts are more properly said to *obtain* than to 'exist'. A serious one is that, since facts are simply what are picked out by true statements, once the notion of fact is rendered conception-dependent, so too is the notion of *truth*. Kuhn may have wanted that, but the idea that what's true, what is the case, the facts, and thus that what *exists* all depend on which paradigms are in place seems to erase any kind of objectivity, by erasing the world itself. One should be able to resist whig historiography, and to present the history of science, without doing *that*.

Scientific revolutions do occur. But how are scientists brought to make the transition from their old paradigm to a new one?[34] Kuhn starts section XII with a critique of philosophies of science which might answer this question by suggesting that a new paradigm is ushered in by being *confirmed* or an old one shown the door by being *falsified*. He finds their description of theses processes unrealistic. So this section considers 'the process that should somehow, in a theory of scientific inquiry, *replace* the confirmation or falsification procedures made familiar by our usual image of science' (p. 8, emphasis added).

Kuhn's critique of naive falsificationism is more pointed than his critique of verificationism. Naive falsificationists think of scientists as routinely testing their preferred theory against observation, or experience, whereas verificationists in the logical empiricist tradition (Kuhn takes Ernest Nagel as his example) try to show how scientists might compare theories in the light of available evidence.

Kuhn's objection to the latter is that they are inappropriately *idealized*: they require a neutral observation language of the kind whose existence he denies, and they fail to recognize that scientists can only compare *available* paradigms against available evidence.

The naive falsificationist, on the other hand, is even further from the truth. Paradigms *do* get tested, Kuhn concedes, but their test is never part of *normal* science, and is *always* a competition, a *comparison* of more than one paradigm with the results of experiment or observation. In fact, as we have already pointed out, Popper himself, being no naive falsificationist, recognized this from the start. However, the objection that Kuhn then puts seems to target *only* naive falsificationism. He points out that not just *any* failure of match between theory and experience could be grounds for rejecting the theory: that would mean rejecting *all* theories. But then falsificationists require some criterion of *severe* failure to fit. If the falsificationist, though, recognizes that tests involve comparing theories with one another to see how well they fit experience, this objection is less easy to apply. The falsificationist then needs, not a criterion of severe failure to fit, but merely some way of deciding which of two theories fits experience *better*, and Kuhn himself admits that such questions can be answered (p. 147).

Kuhn's ultimate answer to the question about how *how* scientists change their paradigms is partly that many of them simply don't. In line with his view that theoretical innovations are usually made by *young* scientists, he follows 'Planck's principle': Max Planck's suggestion that the adherents of the *old* paradigm stick with their views, and just die off, to be replaced by new, younger scientists who subscribe to the new paradigm (pp. 150–1).[35] Paradigm-change is thus partly a feature of scientific *communities* rather than of individual scientists.

Kuhn takes this kind of *tenacity*, often perceived by outsiders to be dogmatism, to be a necessary part of the revolutionary character of scientific research. Sticking to one's paradigm is what makes normal science possible, and normal science is what makes revolutionary science possible. Nevertheless scientific change does occur, so how? For one thing, young people who have just come into a science are not so tightly wedded to its old paradigm. But there also exists the important phenomenon of scientific *conversion*, in which an individual scientist embraces the new paradigm in place of the old one. What are the reasons for such conversions?

Kuhn argues that there is a multiplicity of such reasons: 'Individual scientists embrace a new paradigm for all sorts of reasons and usually for several at once' (p. 152). *Arguments* for the new paradigm certainly have a decisive part to play. The most prominent *claim* on behalf of the new paradigm will usually be that it solves some of the problems that the old paradigm left outstanding, the ones that led it into crisis. Such claims are more likely to be successful if the new paradigm exhibits *quantitative precision*. On top of this, the new paradigm will force the old scientists in the lab next door to take it seriously if it is able demonstrably to *predict new phenomena* that were unsuspected by the old paradigm. (This kind of prediction is made into a requirement on new theories in Popper's falsificationism.) But if it manages to predict *old* phenomena, that is, phenomena which had been observed before the theory that accounts for them, it will also score points.

These considerations are fairly *objective*: one doesn't necessarily have to share the mind-set of the new paradigm to concede that it is issuing in precise predictions of new and familiar phenomena. But Kuhn treats them as relatively superficial. For one thing, claims about the current actual problem-solving power of a new paradigm aren't decisive: 'if a new candidate for paradigm had to be judged from the start by hard-headed people who examined only relative problem-solving ability, the sciences would experience very few major revolutions' (p. 157). Especially in cases where the paradigms are *incommensurable*, arguments about which problems are more important to solve, and about which paradigm is better at solving which problems, are likely to be intractable. For another thing, claims about relative problem-solving power don't really get at the *real* reasons why scientists change their minds. Claims about problem-solving *potential* are more likely to be persuasive.

This is where another group of considerations plays a part in individual conversions. These are considerations which appeal to the scientist's sense of what is *appropriate* or *aesthetic* (pp. 155ff). Kuhn mentions *neatness, simplicity* and *suitability*. He feels that such 'quasi-aesthetic' considerations are likely to be more decisive in the early phases of the conversion of a few important scientists, since adherents of a new candidate for paradigm will not yet have shown that it can solve more than a few problems. But these considerations will not generate a decision-procedure, an 'algorithm' for theory-choice. (Kuhn clearly thinks of them as less objective and less articulable than considerations pertaining to problem-solving power.) In fact, in the case of the few important scientists who convert to the new paradigm (as well as possibly inventing it), *reasons* come second to *faith*. The faith is that the new paradigm will have the problem-solving power that the scientist envisages.

This is undoubtedly why some commentators have interpreted Kuhn as founding the decision-making mechanism of scientific theory-choice on mystical factors. If paradigm debates are 'not really about relative problem-solving ability' (p. 157), but are merely couched in those terms, the question is just how scientists do decide between different ways of doing science 'on faith' (p. 158). Unfortunately, when Kuhn says that the basis for faith in the chosen paradigm 'need not be rational' (ibid.), this suggests a contrast between faith and reason (judgements of existing problem-solving power). However, the faith he has in mind consists in scientists' estimation of a paradigm's future problem-solving *potential*. Kuhn evidently thinks that scientists have ways of estimating such potential which, although personal and not fully articulable, are nevertheless reliable. Saying that this basis for paradigm-choice 'need not be rational' should mean merely that it can't easily be put in the form of rules, not that it is a *mere* leap in the dark.

Unsurprisingly, Feyerabend and his admirers have objected to Kuhn that scientific revolutions need not be as decisive as he portrays them, and that older paradigms could, in fact, be *resuscitated*.

(See, for example, Feyerabend's use of the 'principle of tenacity' in his article in Lakatos and Musgrave 1970, as well as the introduction to Fuller 2000, and chapter 9 of Fuller 2003.) If theories aren't really *refuted* (shown to be false), but merely *overtaken*, surpassed, why shouldn't they be capable of a comeback? Even if the paradigm scientists, at the time of their paradigm's demise, couldn't see any way to solve the problems which plagued it, later scientists might be in a better position to do so. Paradigm-competition may, as Kuhn says, be a decisive process in social terms, but if paradigms are merely exhausted in terms of their *current* problem-solving resources, rather than falsified, there is nothing intrinsic to them which ensures that they cannot come back from the grave.

It might seem that Kuhn's reluctance to appeal to the concept of *truth* is a liability in this respect. If a theory or paradigm could be the kind of thing that could have been true, then it must also be the kind of thing which could be shown to be false. However, if by 'paradigm' we mean disciplinary matrix, there is no sense to the supposition that a paradigm can be either true or false. Kuhn will have to insist that normal science ensures that a paradigm's resources for solving problems are *exhausted* before the transition to a new paradigm is considered, and that this estimate of exhaustion is not merely provisional.

Study questions

1. Is it true that scientists themselves, even the ones involved in scientific revolutions, don't perceive those revolutions? If so, why? Are scientists worse off than non-scientists in this respect?

2. Can intellectual fields, such as science, be characterized in terms of the kinds of texts they involve? Would science be improved if science textbooks weren't 'whiggish'?

3. Is it true that scientific concepts cannot really be invented? Are facts about natural phenomena the kind of things which can be created, and destroyed, by scientific revolutions?

4. What would be wrong in thinking of puzzle-solving *as* paradigm-testing? Why should there be a time during which a science's paradigms remain untested? If paradigms can compete, and thus be compared with one another in certain respects, why can't scientists compare them with the results of observation and experiment? How might verificationists, and Popperians, respond to Kuhn's critiques?

5. Does Kuhn's list of the characteristics of scientific communities suffice to set such communities apart from all other kinds of professional groups? Or was Feyerabend right to complain that gangs of organized criminals, for example, might well share such characteristics?
6. Might scientists have reasonable ways of estimating the *promise* of new paradigms? Should 'aesthetic' factors in paradigm-competition be thought of as 'subjective'?

SECTION 8: INCOMMENSURABILITY, AND SCIENTIFIC PROGRESS

Paradigm-competition is complicated by the phenomenon of incommensurability, a way in which competing paradigms fail to match one another, which Kuhn factors into its component parts. He concludes that to the (limited) extent that scientists do change their paradigms, they do so in a way resembling conversion. Section XIII, which Kuhn seems to have considered his book's most provisional, considers how the idea of development through scientific revolutions, as previously described, is compatible with the idea that science makes progress. Normal science makes progress in a relatively unproblematic way. But why should science progress through revolutions? Is this just a concomitant of the fact that 'the winners write the history books'? Kuhn thinks not. But he insists that our usual conception of what scientific progress consists in must be revised.

In *The Logic of Scientific Discovery*, Popper remarked in passing that

> If some day it should no longer be possible for scientific observers to reach agreement about basic statements this would amount to a failure of language as a means of communication. It would amount to a new 'Babel of Tongues': scientific discovery would be reduced to absurdity. In this new Babel, the soaring edifice of science would soon lie in ruins. (Popper 1959, p. 104)

But, as we shall now see, revolutionary periods, according to Kuhn, have something very like this character.

Choosing between paradigms is *hard*, because the proponents of competing paradigms, we're told, 'are always at least slightly at cross-purposes' (p. 148). Their being so is a matter of *incommensurability*.

Kuhn explicitly traced this concept back to mathematics, where the term refers to the relationship between two integers whose magnitudes *a* and *b* cannot be represented exactly by the ratio *a/b* (the ratio between the side of a unit square and its diagonal, for example). He later called it a 'metaphor' (RSS, p. 298).

Despite having been the focus of an immense amount of debate in philosophy of science, the terms 'incommensurable' and 'incommensurability' occur only eight times in SSR, and three more in its Postscript. Its first anticipatory mention applies to 'ways of seeing the world' (p. 4), two further occurrences apply to pre-and postrevolutionary normal scientific traditions (pp. 103, 148), one to 'research worlds' (p. 112), one to standards (p. 149), two to competing paradigms (pp. 150, 157), and one to problem-solutions (p. 165). In the Postscript, incommensurability is twice said to be a feature of 'viewpoints' (pp. 175, 200), and once to be a feature of theories (p. 198). The important point here is perhaps that none of Kuhn's uses of the concept in SSR apply to *exemplars*. Incommensurability, in SSR at least, is a relationship between disciplinary matrices and the research 'worlds' they generate. The complexity of such matrices explains why Kuhn can factor the concept into the following elements.

(1) Incommensurability of *problems* (p. 148). Here the idea is that what one paradigm takes as a problem doesn't have to be taken by another paradigm as such. The standard logical empiricist image of science and the Popperian programme concur in requiring that a new theory solve *all* the problems of the older theory it is being reduced to, or replacing. Kuhn tells us that this requirement isn't, as a matter of historical fact, met, since there are important historical examples of what we now call 'Kuhn-loss' (see section 5 of this *Guide*).

Kuhn also describes this feature of incommensurability as a difference in the participants' *standards* or *definitions* of science (pp. 103, 148, 149). But this isn't really elaborated upon.

(2) A semantic incommensurability of *language* and *concepts*. 'Within the new paradigm,' Kuhn says, 'old terms, concepts and experiments fall into new relationships one with the other. The inevitable result is what we must call, though the term is not quite right, a misunderstanding between the two competing schools' (p. 149). This is where the change in the *meanings* of scientific terms makes itself felt. Einstein revised what was meant by the term

'space', just as Copernicus revised what was meant by the terms 'earth' and 'motion'. Had no such change of meaning taken place, the transition from one such viewpoint to another would have been a mere change in *belief*, and adherents of the older viewpoint could rightly have been convicted of having been *mistaken*. But the fact that the meanings of the terms in question did change, although it in no way reinstates their viewpoint, does mean that they were not *merely* in error.

Hoyningen-Huene (1993, p. 210) quite rightly points out that diagnosis of meaning-change relies on there being some distinction between attributes that are *essential* and attributes that are *accidental* to a concept. A change in the former, but not in the latter, will amount to a change in meaning, as opposed to a mere change in belief. Kuhn, under the influence of Quine, officially eschewed the analytic/synthetic distinction which the logical positivists and logical empiricists had leant so hard upon. But he needs *some* such distinction here, and later came to recognize as much (ET, p. 304, note 14).

(3) The third and last aspect of incommensurability, which Kuhn takes to be the most fundamental, is the incommensurability of scientific *worlds* (p. 150). (We dealt with these in section 6.)

In the aftermath of SSR, Kuhn came to feel that philosophers, in particular, had seriously misunderstood the idea of incommensurability. One such misinterpretation, discussed in the Postscript (pp. 198–9), is to think of incommensurability as entirely precluding *communication* between the scientists in question. (Critics were able to come up with plenty of examples of scientists from different normal-scientific traditions who communicated across their divide.) Another (perhaps the most popular) is to think of the relationship in question as precluding the *comparison* of its relata. These readings of the incommensurability thesis just have to be wrong, though, since they would make scientific theory-choice fundamentally *irrational*, whereas Kuhn insists that it isn't. Incommensurability does preclude the sort of *point-by-point* comparison which the usual view of scientific theories implies, but not *holistic* comparison.

Kuhn is quite right to suggest that theory- or paradigm-choice can be rational even in the face of incommensurability. He does so by making a move that has more recently been made in other areas of philosophy. Previous philosophers of science had tied rationality to *rules.* They conceived of a decision as rational if and only if one could justify it by reference to a general rule about the factors in question. Kuhn relates rationality to *reasons*, and thereby to being *reasonable*, rather than to *rules.* This is partly because, in line with his thesis of the priority of paradigms over rules, he thinks there simply are no rules which each individual scientist in a scientific community can appeal to *and* which would yield a clear and univocal decision in each case of theory- or paradigm-choice. Nevertheless, Kuhn insists that there are good *reasons* why scientists choose one theory or paradigm over another. That is, in each case, a choice can be properly justified by reference to individual factors, but not necessarily by reference to any *general* rule about how to choose between theories or paradigms. I take it that this is what he means by saying that debates over theory-choice cannot be decided by any algorithm or cast in the form of proof. Nevertheless, reasons operate within these debates, and they involve reference to features such as accuracy, simplicity and fruitfulness. These features function 'as values' (p. 199), rather than as rules, in that although different scientists may make reference to the *same* features, they can be differently applied.

The ultimate decision, though, is made not by the individual scientists in question, but by the relevant scientific *community*. This is how Kuhn seeks to register a kind of *democracy* which must operate within science: no single scientist, however powerful, controls a scientific community. A single scientist may influence his or her peers, but those individual peers each have to make up their own minds about which theory to subscribe to.

However, this idea that it is the scientific community that decides between theories or paradigms has led some critics, such as Tim McGrew, to accuse Kuhn of sliding between different senses of 'scientific community':

The point is simply that we trust scientists because we are convinced that they are acting in a scientific fashion, and this is not defined as what large numbers of people who call themselves scientists decide to do. When we discover that this assumption has played us false – that the scientists are being unscientific – even well-informed laymen can make the relevant discriminations. Kuhn's use of the term 'scientific group' is perfectly ambiguous between these two readings; if we take it to refer merely to self-styled scientists, then the obvious response is that we must discriminate between self-nominated prophets of science; if we take it to refer to those who are truly scientific, then it presupposes what Kuhn is apparently trying to deny: that it is possible to identify the truly scientific community in a way that is independent of appeal to that community. (McGrew 1994, p. 4)

As we shall see below, I don't think Kuhn is open to this objection, for I don't think he is denying that we can identify scientific communities independently of whether members of a community *say* they are scientists. Kuhn clearly takes a community to be a *scientific* community only if its members are responsive, in their decision-making, only to certain kinds of consideration.

Nevertheless there is still a problem with Kuhn's notion of incommensurability. The original, mathematical notion is of a relation between *objects* (viz., numbers), rather than between normal-scientific traditions. But with representations and concepts there is an extra dimension to the *relata*: the objects have to be representations or concepts *of* a common something, a common domain. (If this condition isn't met, the claim of incommensurability reduces to a *triviality*. It's no news at all, for example, that quantum mechanics and evolutionary biology are 'incommensurable': they share no subject-matter.) But the existence of this common target or domain seems to belie the claim of incommensurability itself, since it suggests that there *must* be a 'neutral' way to identify what the two supposedly incommensurable representations are about, what they are conceptualizations *of*. If this is right, there is at least *some* neutral 'language', and Kuhn's claim that historians cannot find one in cases of incommensurability is superficial, as well as self-refuting.[36]

Might Kuhn respond that the terms in which the common target or domain is specified need *not* be 'neutral'? What he says about the way scientific fields change over history (e.g. RSS, pp. 290, 295) might suggest this. Could our specification of the common domain be merely *our* identification of what is common, one the older scientists might not or could not share? The problem here is that if the way we identify the common domain is merely *our* way of doing so, we would seem to have fallen short, by Kuhn's standards, in respect of neutrality between our theory and theirs. We would seem to have taken sides, something which Kuhn's kind of historiography officially refuses to let us do.

Another tempting response might be that although there's *some* 'neutral' language, there is simply not enough to form a basis on which to compare the two rival paradigms. However, I suspect that this is where the power of one of the logical positivists' ideas reveals itself. The early logical positivists, in particular, were acquainted with an early version of the incommensurability thesis (due to the followers of Henri Poincaré's 'conventionalism', notably Edouard LeRoy and Kasimierz Ajdukiewicz). Against it they proposed their thesis of '*physicalism*', according to which a certain 'language of physical objects' constitutes the intersubjective language of science. The 'language' in question, a refined subset of the ordinary languages all of us speak, is one in which physical objects and phenomena can be picked out by primitive spatio-temporal descriptions. Scientists who disagree about which heavenly bodies are planets, for example, can at least agree about which points of light in the night sky should be observable at different times. In saying that this kind of language was the language of 'unified science', one thing the positivists meant was that it was, or could be, a basis on which different scientific theories could be commensurated.

Of all the concepts forged in SSR, incommensurability was the one that Kuhn worked on the most in the years that followed. In the book's Postscript, he makes the first of several changes to its presentation, by reconceiving incommensurability in more explicitly *linguistic* terms.[37] Scientists who have incommensurable 'viewpoints' are now to be thought of as 'members of different language communities' (p. 175), and their speaking from those different viewpoints is a matter of their 'using words differently' (p. 200).

Recall that sections IX to XIII of SSR were supposed to demonstrate the parallels between science and politics revealed by historical study (p. 94). And the central such parallel was of course that paradigm-change is a matter of *conversion*. Kuhn now introduces an important distinction between conversion and *persuasion*, allotting a significant (but limited) role to the latter.

Unfortunately, this subsection is one of the most difficult of his Postscript. Things begin promisingly enough, with Kuhn explaining incommensurability in terms of a change in basic similarity relations:

> The practice of normal science depends on the ability, acquired from exemplars, to group objects and situations into similarity

sets which are *primitive* in the sense that the grouping is done without an answer to the question 'Similar with respect to what?'. One central aspect of any revolution is, then, that some of the similarity relations change. Objects that were grouped in the same set before are grouped in different ones afterward and vice versa. (p. 200, emphasis added)

Scientists who communicated with one another easily before such a regrouping may find that they can no longer do so afterwards. There is no single 'neutral' language in which their rival theories can both be stated. What such scientists still *do* share, though, affords them a way of re-establishing contact with one another, even though the means in question involve the kind of interpretation historians of science engage in, rather than anything familiar from natural science itself.

There are several aspects (or perhaps stages) of this *rapprochement*. Because he now explicitly conceives of incommensurability as *local*, rather than global, Kuhn recognizes that even scientists with incommensurable theories can try to persuade one another of the virtues of their own theory by resorting to those parts of their theories in which terms are used by both groups in the same way. This can afford a partial comparison of their theories' problem-solving ability, which may be sufficient to persuade some scientists, particularly young ones, of the superior merit of their rivals' theory (p. 203).[38]

Kuhn, though, doesn't think this effect will be extensive or decisive. In line with his linguistic conception, a second aspect of incommensurability involves the adherents of each theory recognizing that they now form different language communities, and then having recourse to *translation* (pp. 202–3). This process affords understanding of, yet without commitment to, the other group's theory. In a third and final stage, though, scientists from one group may 'go native', making the theory held by the other group their own. This process, Kuhn insists, inevitably involves not persuasion, but the kind of *conversion* upon which SSR focused.

Kuhn didn't doubt that science *progresses*. But he wanted to know *why* and *how* it does so in ways that other disciplines don't. Having already deprived himself of the 'usual' answer to these questions (in terms of cumulative growth in knowledge), he had to find others. And, having stressed the existence of scientific *revolutions*, his answers have to be compatible with their existence.

Part of his answer is that our concept of progress isn't *independent* of our concept of science (just as it isn't independent of our concept of technology). Practitioners of fields whose status as scientific is uncertain shouldn't think of themselves as needing a 'definition' of science. Our concept of science *already* dictates that sciences make progress, and Kuhn thinks it clear that some fields don't obviously progress in the way that physics, biology, etc. do. The real question is *why* they don't do so. And his answer to this question is that it's *consensus* about accomplishments which facilitates progress, and thus makes a field scientific.

Kuhn warns us that in coming to see things aright, we will have to 'invert' our normal view of the relation between science and the community that practises it (p. 162). He also warns us that when we do so, seeing as causes what have been generally taken as effects, the phrases 'scientific objectivity' and 'scientific progress' 'may come to seem in part redundant' (ibid.). By 'redundant' here he presumably means *pleonastic*: our concepts of objectivity and of progress are implicated in our concept of science, and vice versa.

Kuhn treats the question of the presence of progress in a divided way, depending on whether what's in question is progress during *normal* science (pp. 162–6), or progress across scientific revolutions (pp. 166–73).

Normal science progresses because normal scientists do success-ful creative work, adding to the collective achievement of the group. In this respect they resemble the members of any other profession. By contrast, non-scientific fields fail to exhibit progress not because their individual schools fail to progress, but rather because they are characterized by *competing* schools, 'each of which constantly ques-tions the very foundations of the others' (p. 163). Of course, as we saw earlier, pre-paradigm periods of scientific activity, together with periods of extraordinary scientific research, are the same in this respect. So only during normal science, under the hegemony of a single paradigm, is cumulative progress both ensured *and* obvious (its obviousness being why it is registered in the 'usual' image of science). During such periods, the concentration of atten-tion upon esoteric problems, and the isolation of mature scientific communities from everyday life and demands, help to increase the effectiveness of the group's problem-solving efficiency. Problems themselves are selected because they look as if they might be solvable, rather than because of their social importance. The narrowness, specialization, rigidity and unhistorical nature of scientific training helps to ensure that solvable problems are identified, and then solved. This training isn't designed to produce new ideas, and that's why new ideas tend to come from new thinkers. But this *rigidity* of the individual scientist's training and conduct is wholly compatible with the *flexibility* of the community, it's being able to move from one paradigm to another.

Foundational issues, though, *aren't* dealt with during such periods. If they were, and if they really had an effect upon the course of the work undertaken, that effect could only be retarding. Scientists make progress because they choose their own problems and, most of the time, they don't allow themselves to get mired in philosophy.

So there's an unproblematic sense in which normal science progresses, and quite objectively so: puzzles get solved. The *problem* about scientific progress concerns progress through scientific revolutions themselves, progress from one paradigm to another. Why should progress accompany revolutions? Is this just a matter of definition? Here Kuhn acquiesces:

> Revolutions close with a total victory for one of the two opposing camps. Will that group ever say that the result of its victory has been something less than progress? That would be rather like admitting that they had been wrong and their opponents right. To them, at least, the outcome of revolution must be progress, and they are in an excellent position to make certain that future members of their community will see past history in the same way. (p. 166)

The perception of progress *within* science, then, is inevitable. Its practitioners must come to see the history of their subject as a direct trajectory to their present position. They must be implicated in 'whig historiography'. But Kuhn has already told us (in section XI) that this kind of historiography embodies the *illusion* of cumulative progress. So is it that might makes right in the historiography of science?

Kuhn emphatically says no. Scientific communities can only take part in *scientific* revolutions when it is professional *scientific* community authority that forces the transition: 'The very existence of science depends upon vesting the power to choose between paradigms in the members of a special kind of community' (p. 167). And these scientific communities are a historically localized phenomenon, arising only in Europe, and only in the past four centuries. No other civilizations have established them.

This raises the question 'What *are* the special characteristics of these scientific communities?' Kuhn lists several, but only in a provisional way. Some concern the attitude of the *individual* scientist (s/he must be concerned to solve problems about the behaviour of nature, the problems on which s/he works must be problems of detail, the solutions that satisfy her or him must be accepted by others as solutions); others concern the social nature of the *group* (the group that shares the problems must be 'the well-defined community of the scientist's professional compeers' (p. 168), the appeal to any kind of external authority to decide scientific questions must not be acceptable). In Kuhn's Postscript these characteristics are listed in a different way. Progress in the mature sciences is there said to be connected with 'the relative scarcity of competing schools' in science, 'the extent to which the members of a given scientific community provide the only audience and the only judges of that community's work', 'the special nature of scientific education', the goal of puzzle-solving, and 'the value system which the scientific group deploys in periods of crisis and decision' (p. 209).

Kuhn certainly shared the general worry that characterizing science in terms of communities and the way people are instructed within them leaves the door open to its perversion. Cavell recalls a conversation in which Kuhn said, 'If instruction and agreement are of the essence of the matter, then Hitler could instruct me that a theory is true and get me to agree.' Cavell's response, with which Kuhn came to agree, was to deny that a dictator could *convince* one of, or *educate* one in, the truth of a theory: 'Hitler could declare a theory to be true, as an edict. He could effectively threaten to kill you if you refuse to, or fail to, believe it. But all that means is that he is going to kill you, or perhaps kill you if you do not convince him that you accept and follow the edict.'[39] A scientific community is a community whose members share certain things (paradigms, theories, beliefs). But they must have come to them in a special kind of uncoerced way, a way responsive to the evidence in question, and the paradigm's promise in dealing with it. Kuhn was not particularly concerned with contemporary 'big science', but his stance here is particularly relevant to the issue of whether *commercial* interests now warp the decision-making of certain erstwhile scientific communities.

A community that shares the special characteristics Kuhn notes *must* see their own paradigm-change as progress. Such communities, after all, are designed to maximize 'the number and precision of the problem[s] solved through paradigm change' (p. 169). If they switch to a paradigm without that quality, they are no longer doing their job. So their perception of progress is, as Kuhn puts it, 'self-fulfilling'.

Nothing sinister was meant by this. Rather, Kuhn was pointing out that the community in question is the very best kind of potential authority on whether a paradigm-change *is* progressive. No-one from outside the community in question could be in a better position to make the decision. This is because no-one could be in a better position to assess whether the conditions on a new candidate for paradigm are fulfilled. The two conditions in question reflect the two aspects of paradigm-choice: one backward-looking and one forward-looking. (They also mirror the two reasons why paradigms gain their adherents in the first place, during the pre-consensus period.) The first condition is that the candidate must resolve an outstanding and generally recognized problem. But this isn't a 'unique or an unequivocal basis for paradigm-choice' (ibid.). The second such condition is that it must promise to preserve the concrete problem-solving ability of past science. A scientific community cannot switch to a paradigm that would leave them in a worse position, in terms of problem-solving, than their existing paradigm. This helps explain why so many pictures of science are cumulativist: scientists will only indulge in scientific revolutions when a *better* paradigm is in the offing.

There is, however, a problem lurking here (as Meiland 1974 and McGrew 1994 have noted). Can the puzzle-solutions of a new paradigm be compared with those of an old paradigm that is *incommensurable* with it? It is hard to see how this can be so if the old puzzles cannot even be formulated within the new paradigm.

Finally, at the very end of the first edition of SSR, Kuhn turns (for the first time) to the topic of truth. He does so only to suggest that although science inevitably progresses, we need not acquiesce in the usual image's idea that it progresses *towards the truth*. The developmental process that SSR has described is a process of *evolution*, but (like biological evolution) it is not an evolution *toward* anything, not a teleology. Neither the existence of science nor its success require us to suppose that there is either a *goal* of science, or even that there is 'some one full, objective, true account of nature' (p. 171). And Kuhn suggests that certain philosophical problems will vanish if we *don't* make these suppositions.

Kuhn only really sketches this 'Darwinian analogy' here, outlining its major ideas: that the resolution of revolutions is analogous to a process of selection by conflict of the fittest way to practise science; that the net result of a sequence of such selections is 'the wonderfully adapted set of instruments we call modern scientific knowledge' (p. 172); that successive stages in the process of development are marked by an increase in articulation and specialization; and that the entire process may have occurred without the benefit of a set goal. Kuhn took this Darwinian analogy seriously, and later insisted that it 'should have been taken more seriously than it was' (RSS, p. 307). But most commentators find it lacking in detail, and there are even worries about whether it is compatible with the account of scientific development outlined in SSR's early chapters. (Other philosophers of science have tried to flesh out such analogies, though.)[40]

Kuhn was undoubtedly right that we don't need the idea of teleology to make sense of the idea of evolution in general. But since science, unlike biology, is an intellectual process in which human groups engage, it's still a legitimate question whether a goal *might* be involved, whether science as an institution might have an aim (in the way that medicine might be thought to aim at the eradication or amelioration of illness). In his Postscript, though, Kuhn went further, explicitly denying that truth, conceived of as a theory's correspondence to reality, could constitute such a goal for science. 'One often hears,' he says, 'that successive theories grow ever closer to, or approximate more and more closely to, the truth', meaning that a theory's ontology comes to attain a match 'between the entities with which the theory populates nature and what is "really there"' (p. 206). He found such a notion of truth untenable because there is no theory-independent way to 'reconstruct' phrases such as 'really there', and (thus?) because the idea that a theory's ontology might correspond to its 'real counterpart in nature' is illusory.

Alexander Bird's characterization of these remarks seems to me decisive. Kuhn, he says, must share our intuitive notion of the possibility of error and of ignorance, since the only satisfactory explanation of the origin of *anomalies* is that the world is not exactly as our theories say it is. If error or ignorance can be shared by all of us, then there must be a way things are that is 'beyond' theory. [. . .] Even if it were impossible to assess the assertion of a match, that would not make that assertion *meaningless*, unless one had some sort of verificationist view about meaning. (Bird 2000, p. 227, emphasis added)

I agree with Bird that Kuhn's remarks about the 'foundationalism' he associated with the correspondence conception of truth are indeed verificationist, and that verificationism now bears the onus of proof.

By rejecting this 'correspondence' conception, Kuhn didn't mean to reject the concept of truth per se. In fact, in his later writings he declared that it should be replaced by 'a *strong* conception of truth' (RSS, p. 95, emphasis added), and made two incompatible suggestions for what this might be. The less interesting (which may well have been inspired by reading Hilary Putnam's middle-period works (see RSS, p. 312)) is the concept of rational assertibility. The more interesting (and more Wittgensteinian) suggestion is to replace correspondence truth with 'something like the redundancy theory' (RSS, p. 99). But Kuhn never followed up on this.

His whole approach was an attempt to get away from seeing science as consisting entirely in a system of linguistic representations for which the question of truth-or-falsity can arise. As he once put it:

I take theories to be whole systems, and as such they don't need to be true or false. All we need to do is by some criteria or other decide which one we would rather have. In general, this is roughly specifiable, but that doesn't get me into the true-false game. Of course it doesn't eliminate true-false as very *important*. That's what you do *within* a system, – judge the truth or falsity of statements. *Across* a system you can't apply that sort of calculation. (Sigurdsson 1990, p. 22)

Kuhn's reminder that science includes components which are not evaluable as either true or false is salutary. Denying that science has anything to do with truth is wildly implausible. As MacIntyre once pointed out, science has certainly shown us that some existence-claims are *false*, 'just because the entities in question are *not* really there – whatever *any* theory may say' (MacIntyre 1977, p. 469). But if, as we have seen, paradigms (exemplars, at least) can *be* theories, which for Kuhn in SSR they certainly can, and if theories are or include *claims*, the question of their truth still arises. Theories are not merely 'systems'. Indeed, even if one focuses on problem-solving, as Kuhn does, the question of truth may still be implicit, since solving a problem of *theory* presumably involves giving it an answer that is *true* (or approximately true, or likely to be true). Theories that solve more and more problems *are* theories that say more and more true (and fewer false) things. (There might still be a question whether this means that sciences themselves can be said to move closer and closer to the truth, though.)

In his Postscript, Kuhn continued to insist that science progresses in a special way that marks it off from other disciplines (p. 209), and he took this insistence to constitute a reply to certain accusations of relativism (p. 205). What he said about these issues, although schematic, is what makes me want to compare his views to the idea known as *conceptual* relativism, relativism about concepts.

Kuhn agreed that, when applied to cultural issues, the idea that members of different language-culture communities may both be right is relativistic. But against the accusation that *he* was a relativist, he insisted that when applied to *science* the idea that proponents of different theories may both be right may not be relativism, and that 'it is in any case far from *mere* relativism' (ibid., emphasis added). The pre-eminence accorded in science to the virtue of puzzle-solving, he argued, means that one can sort theories in a given scientific field into earlier and later ones in a paradigm-neutral way. As he put it: 'Later scientific theories are better than earlier ones for solving puzzles in the often quite different environments to which they are applied. That is not a relativist's position, and it displays the sense in which I am a convinced believer in scientific progress' (p. 206).

Kuhn nevertheless continued to resist the idea that science moves closer and closer to the truth. I think one can at least make sense of why he said this. Scientific *theories* may, as I have suggested, be candidates for truth-or-falsity, but scientific *conceptual schemes* are not. Kuhn's larger concern, as we have seen throughout this *Guide*, was primarily with the latter, under the name 'paradigms' (later, 'disciplinary matrices'). He simply did not concern himself much with the issue of truth. He can be taken to be insisting, rightly, that the conceptual schemes scientists devise can be assessed only in prag-matic terms. New paradigms (disciplinary matrices) bring resources which are better than those of their predecessors at solving the the-oretical problems their scientific fields present. This doesn't make conceptual schemes themselves 'better representations of what nature is really like', simply because they are toolkits, not represen-tations. Kuhn was thus right to remind us that they are neither true nor false. Nevertheless, such conceptual schemes are what allow sci-entists to *make* claims, and *these* can still be assessed as true-or-false.[41] Kuhn was wrong if he thought he had to *evade* or ignore the issue of truth: the view underlying what he had to say enables one to put the issue in its proper context. Relativism about truth is unten-able: scientific claims are either true or false. And scientific theories are or involve claims, they are not merely 'systems'.

None of this, though, means that natural-scientific phenomena can be adequately conceptualized only in terms of *one* conceptual scheme (as determined 'metaphysical realists' and 'scientific realists' seem to think). There need not be only one 'way the world is'. Claims about the same phenomena, made within conceptual schemes whose concepts categorize the domain of those phenomena in incompati-ble ways can, in principle at least, be true together. If incommensu-rability makes sense, claims made within conceptual schemes whose concepts categorize a single domain of phenomena in incommensu-rable ways may even be incommensurable with one another *and* true.

Study questions

1. Is paradigm-competition as difficult as Kuhn makes it seem? Which aspects of incommensurability are the most compelling? Is Kuhn's view of the resolution of paradigm-competition com-patible with the idea that certain scientists, at least, can hold more than one competing paradigm in their minds at a time?

2. Do scientists really *choose* between theories, or between para-digms, on Kuhn's view? Given his subscription to Planck's prin-ciple, can Kuhn really think that scientific communities make such choices in a democratic way? Even though not all members of the community do choose? And even though those who do choose may not be representative? Isn't 'Planck's principle' straightforwardly empirical, falsifiable and falsified?

3. Is Kuhn asking the *conceptual* question 'What does scientific progress consist in?' or the *causal* question 'What makes science

progress?' Do his qualifications of the way science progresses through revolutions open him up to the charge of relativism? Having held the historian *apart* from the scientific communities, as Kuhn wants us to, isn't there, as well as the question of whether such communities perceive progress, also still a question of whether progress has actually been made?

4. Does Kuhn operate a double-standard in insisting on truth in the history of science, but denying any notion of truth to natural science? If, as Kuhn suggests, the concepts of objectivity and progress are connected with our concept of science, why isn't the concept of truth, or of movement towards the truth, also so connected?

5. Is verificationism implicit in Kuhn's suggestion that talk of what exists outside of theory is meaningless?

CHAPTER 4

RECEPTION AND INFLUENCE

SSR was critiqued even before it was published. Kuhn's Berkeley colleague Paul Feyerabend wrote important letters to him in the early 1960s (published in Hoyningen-Huene 1995 and 2006). Soon after having completed the first draft of SSR (but before the book's publication), Kuhn abstracted from it a paper entitled 'The Function of Dogma in Scientific Research' (published in Crombie 1963), which he later came to think of as unsatisfactory. Feyerabend wrote an important review of this volume (Feyerabend 1964) in which he took issue with Kuhn's paper. On the relationship between Kuhn and Feyerabend, see their essays in Lakatos and Musgrave 1970, plus Preston 1997 (especially chapter 5) and Hoyningen-Huene 2000. Feyerabend doubted the historical accuracy of Kuhn's picture of science, and strongly criticized both the idea that the restrictive aspects of normal science are 'functional' and Kuhn's presentation of his picture, which Feyerabend considered an illicit mixture of the descriptive and the normative. Feyerabend's own works (e.g. Feyerabend 1975, 1978 and the earlier papers now collected in Feyerabend 1981a, 1981b, and 1999) constitute an important and explicitly 'pluralistic' alternative to Kuhn's perspective.

Within the first tranche of secondary literature, most early reviews of SSR were positive (see, for example, Hesse 1963; Bohm 1964), and there were important reactions from sympathetic figures (e.g. Michael Polanyi's comments in the 1963 Crombie volume). But Israel Scheffler (1967, 1972, 1982) vigorously defended the logical empiricist perspective against Kuhn's critique, and many philosophers of science were quite dismissive.

In 1965, Kuhn was drawn into a famous debate and confrontation with Karl Popper and his principal supporters at the Bedford colloquium, in London. Its proceedings were published as Lakatos and Musgrave 1970. But Kuhn did not take all the papers to be hostile; indeed he acknowledged that Margaret Masterman's had helped him clarify the concept of a paradigm.

Dudley Shapere, an original but rather neglected philosopher of science, published important reviews of the first and second editions of SSR (Shapere 1964, 1971), as well as an extended critique of Kuhn's and Feyerabend's work (Shapere 1966).

Another, even more important and neglected voice is that of Stephen Toulmin. Having published important works that antici-pated SSR in certain respects, he reacted badly to the book (e.g. his essay in Lakatos and Musgrave 1970). The evolutionary perspective on scientific change that he was already developing, though (Toulmin 1972), was better worked-out than the evolutionary aspects of Kuhn's, and a serious rival to its other, revolutionary aspects. Kuhn felt that he and Toulmin, although on personable terms when they initially met during Kuhn's trip to England in 1950, never got on since Toulmin came to the US in 1965 (RSS, p. 297).

Many of these critics, despite their rather different views and backgrounds, levelled at Kuhn the same accusations of irrational-ism, relativism and idealism. These accusations exasperated him, and he did not always have the patience to reply to them in detail. But several of the papers in ET and RSS do constitute such replies.

These same charges, though, were taken by other audiences as compliments. Irrationalism and relativism, in particular, although anathema to most philosophers, had great appeal to certain social scientists, as well as to Kuhn's wider lay audience. Among certain groups of sociologists of science, the idea that Kuhn had played a major part in debunking natural science took hold. In Britain, Kuhn's work fed back into the tradition of sociology of scientific knowledge, to take its place alongside the work of Karl Mannheim and Ludwik Fleck in the foundation of the 'Edinburgh School', which flourished during the 1970s and 1980s. The school's central figures, David Bloor and Barry Barnes, tried to push Kuhn's work in a more sociological and 'externalist' direction which Kuhn himself resisted. Outside academia, the very terms of SSR's title, particularly the idea of revolution, appealed to an audience of student radicals that Kuhn did not court. Indeed, he took pains to distance himself from all these groups of followers, considering their appropriations of his work to be just as much misinterpretations as those prevalent within the philosophy of science community. The historiographic revolution, then, did not really go the way Kuhn anticipated, and he came to play the role of an elder whose approach was more often than not honoured in being breached.

Among philosophers of science, SSR was far more influential in weakening the dominant logical empiricist approach to science than in replacing it with any single alternative. In the wider intellectual context of the humanities and social sciences, though, Kuhn's book was perceived as a death-blow to positivist, empiricist *and* 'realist' philosophies of science, and it did usher in a vague new 'post-modernist' perspective on science which seems to persist there, whatever the recognized problems with Kuhn's own approach.

SSR, then, provoked strong reactions among philosophers of science. Kuhn anticipated that once its new image of science 'was elaborated in more detail, and once the factors lying behind it had been investigated more thoroughly, ways of delineating it would emerge that would be less open to philosophic complaint' (Buchwald and Smith 1997, p. 368). He did indeed spend much of the rest of his academic career exploring different ways of elaborating and defending ideas from SSR. But the grand revision of that work which he anticipated never appeared. Instead, Kuhn pursued specific ideas such as meaning-change and incommensurability, reworking them in ways that would continue to make contact with the history of science, while avoiding philosophical objections. Kuhn (prefigured here by Ryle and Polanyi) was originally one of those who moved away from treating science as an entirely *linguistic* structure. But although SSR was an improvement on existing views of science in characterizing science as skilled *activity*, not just *statements*, *practice*, not just *theory*, Kuhn later showed signs of going back on this. In his later work, he thought he had 'messed up' the concept of incommensurability by associating it with a change of *values*, rather than *directly* with language (RSS, p. 298). He later seemed to think he hadn't talked enough about meaning-change in SSR (ibid.). He insisted that those 'historical philosophers of science' who dropped the problem of meaning dropped incommensurability, and therefore had eliminated what for him was the philosophical problematic (RSS, pp. 309–10). Here, I agree with Hacking, who complains about Kuhn's too-exclusive focus on meaning and language. Hacking himself was one of those who followed up Kuhn's work in a more profitable way by developing what has come to be called the 'new experimentalism' (see Hacking 1983), a renewed focus on the experimental aspect of natural-scientific activity.

It was certainly as a philosopher, rather than as a historian, that Kuhn had the most influence. Very few came to *do* history in the way SSR suggested, and in his later work he narrowed his focus to concentrate on SSR's philosophical aspects.

In the 1970s Kuhn's work attracted a new band of philosophical converts, occasionally known as the 'new Kuhnians'. Gerald Doppelt's 1978 article is the best-known product, but certain works of Harold I. Brown (e.g. Brown 1979, 1983) and Jack Meiland (e.g. Meiland 1974) also deserve to be included. It is difficult to say what impact they had, but they certainly kept alive the Kuhnian perspective within the philosophy of science, although Harvey Siegel frequently brought their work under critical fire.

There has certainly been a revival of interest in Kuhn in recent years, following the publication of Paul Hoyningen-Huene's magisterial 1993 book, and the third edition of SSR (1996). In today's more diverse philosophical climate, critical appraisals and defences have come from very different directions.

Alan Sokal and Jean Bricmont (1998), for example, linked Kuhn with certain disreputable tendencies in recent Continental European thought, and attacked what they think of as his epistemological relativism. This latter charge has been replied to, though, in Glock 2000.

Michael Friedman's work (1999, 2001) invokes Kuhn (alongside the logical positivists, logical empiricists and Kant) in a synthesis which deserves serious consideration.

Paul Thagard is one of several philosophers of science influenced by Kuhn who have tried to pursue his approach in a way that makes contact with contemporary cognitive science (see Thagard 1992). Ronald Giere is another. Along similar lines, Alexander Bird (2000) has attempted to assimilate Kuhn to contemporary philosophical 'naturalism', bringing out the ways in which Kuhn anticipated and sympathized with the idea that certain philosophical problems about science might be addressed with the tools of science itself.

Steve Fuller's fascinating but controversial book *Thomas Kuhn: A Philosophical History for our Times* (2000), is both a study of the context of Kuhn's work, and a work of 'reception history', presenting Kuhn as a cold-war conservative. Many articles devoted to Fuller's reading of Kuhn are collected in the volume 17, 2003 issue of *Social Epistemology*, edited by Stefano Gattei.

For Wes Sharrock and Rupert Read (2002), *pace* both Bird and Fuller, Kuhn is a philosophic radical, following Wittgenstein (to a certain extent) in attempting to offer therapies to counter philosophical misconceptions about science.

Many contemporary philosophers of science, including most of those best thought of in academic circles, have taken on aspects of Kuhn's work. One of the most important and prevalent reactions, though, has been to try to marry elements of Kuhn's approach with the dominant concern in contemporary philosophy of science, which has been the realism/anti-realism debate, and in particular with the dominant *option*, which is scientific realism. Kuhn's project, one might say, has been very influential, but the particular options which he seems to have taken have not proved as attractive, and the concern with truth, together with a metaphysical concern with ontology, have reasserted themselves in ways that Kuhn would not have been much interested in. Likewise, the general cultural climate of *scientism*, the supposition that science can address all our questions, and is the best model of all intellectual activity, is something strongly opposed in Kuhn's work, but also something which that work could hardly have succeeded in dispelling.

Kuhn's legacy, therefore, is rich, but divided. One of his most illustrious former graduate students, John Heilbron, summed up Kuhn's career as follows:

Although he had few doctoral students in history and none on philosophy, he had an immense readership; no true disciples, but a worldwide congregation. He transformed his contemporaries' understanding of the nature of science and changed the world for those who study the problems that concerned him. His achievement is not easy to explain. He drifted from one academic field to another; his formal equipment for historical research was rudimentary; *Structure* is full of holes; *Black-Body Radiation* is impenetrable; the big book on philosophy has not appeared. (Heilbron 1998, p. 514)

CHAPTER 5

GUIDE TO FURTHER READING

The three best monographs on Kuhn are Hoyningen-Huene 1993, Bird 2000 and Sharrock and Read 2002. The first, a detailed commentary on SSR and a reconstruction and defence of Kuhn as a systematic Kantian philosopher, is still the best book on the subject, and had Kuhn's *imprimatur*. It also features a far more extensive bibliography of works on Kuhn than can be presented here. Alexander Bird's book brilliantly develops the more naturalistic aspects of Kuhn's work, whereas Wes Sharrock and Rupert Read's plays these down in favour of presenting Kuhn as a therapeutic Wittgensteinian. (Similar Wittgensteinian takes are outlined in Robinson 1996, and Kindi 1995.)

Good short introductions to Kuhn's work as a whole include Andersen 2001, von Dietze 2001 and Marcum 2005. Von Dietze's book approaches Kuhn from the perspective of science education.

Lakatos and Musgrave 1970, Gutting 1980, Horwich 1993 and Nickles 2002 are all useful collections of essays on Kuhn. Cedarbaum 1983 is still one of the best articles written on the concept of a paradigm in general. I. B. Cohen's 1985 book gives a detailed historical treatment of the idea of scientific revolutions, but Ian Hacking's 1986 review of it is also important. Suppe 1977 is an important confrontation between the 'received view' of scientific theories and the view that then appeared to be emerging from the new 'historical' philosophy of science.

For information on Kuhn's life, see Buchwald and Smith 1997, Heilbron 1998, Andresen 1999, Marcum 2005, the interviews Kuhn gave in RSS and Sigurdsson 1990, and Keay Davidson's forthcoming biography of Kuhn.

Fourteen papers Kuhn wrote between 1957 and 1974 were

reprinted in *The Essential Tension: Selected Studies in Scientific Tradition and Change* (Chicago: University of Chicago Press, 1977). Hacking's 1979 review of this volume raises several important questions about the nature of Kuhn's work. Despite being critical of Kuhn in many respects, and disavowing several of his key terms, Hacking has developed a view which marries aspects of Kuhn's work with aspects of the work of Michel Foucault and the British historian of science A. C. Crombie (see, for example, Hacking 1985). But Hacking prefers Kuhn's *earlier* papers to SSR.

Brendan Larvor's 2003 article is excellent on the very significant influence historicist thinkers such as Koyré and Butterfield had on Kuhn.

A recent issue of the journal *Tradition and Discovery* (vol. 33, no. 2, 2006–7) was dedicated to the relation between Kuhn's work and that of Michael Polanyi. (Even if Kuhn misunderstood Polanyi in certain respects, he does seem to have taken over certain ideas from him.)

On the relation between Kuhn's work and that of the logical empiricists, especially Rudolf Carnap, see Reisch 1991, Axtell 1993, John Earman's essay in Horwich 1993, Irzik and Grunberg 1995 and several works by Michael Friedman, e.g. the papers collected in Friedman 1999, Friedman 2001 and his paper in Nickles 2002. The consensus seems to be not only that Carnap's *later* views are far closer to Kuhn's than Kuhn recognized, but also that Kuhn deceived himself in thinking that he had really joined issue with logical empiricism. (For a dissenting voice, though, see Pinto de Oliveira 2007). The same verdict arises from works (e.g. Bird 2000, 2002, Gattei forthcoming) portraying Kuhn as having inherited important but unacceptable features of logical positivism and logical empiricism. I attempted to defend Kuhn from some aspects of this charge in Preston 2004. Bird replied in his 2004 article.

The very idea of 'conceptual relativism' was attacked, famously, in Davidson 1974, but was put decisively back on the agenda, in my view, by Hacker 1996, and has also been explored in Arrington 1989 and Glock 2007. Unlike Kuhn, whose proper concern was only with *scientific* conceptual schemes, Davidson's concern was with *total* conceptual schemes.

The accusation of irrationalism was levelled at Kuhn by various philosophers, among whom one of the more worthwhile was David Stove (Stove 1982).

Of all the things emerging from the new philosophy of science of the 1960s, the concept of incommensurability, developed in parallel by Kuhn and Feyerabend, has had the most attention devoted to it. Wisdom 1974, Szumilewicz 1977, Devitt 1979, Moberg 1979, Musgrave 1979, Short 1980, Grandy 1983, Kitcher 1983, Burian 1984, Collier 1984, Franklin 1984, Hoyningen-Huene et al. 1996, Siegel 1987, Wong 1989, Biagioli 1990, Hacking's essay in Horwich 1993, Malone 1993, Kindi 1994 and 1995, Sankey 1994 and 1997, Hoyningen and Sankey 2001, and Jacobs 2002 are all relevant.

Janet Kourany's 1979 paper raises significant problems for Kuhn's claims that scientific development isn't cumulative on the theoretical or on the factual level, and their supposed justification by historical sources. Along the same lines, the important but rather neglected 1988 volume edited by Donovan, Laudan and Laudan features papers assessing the empirical credentials of several of the 'new' historically oriented philosophies of science, including Kuhn's.

Kuhn's later work, and in particular the ways in which he developed the notion of incommensurability, are covered in Chen 1997, Sankey 1997, Irzik and Grunberg 1998, Bird 2000, Buchwald and Smith 2001, Sharrock and Read 2002 and Gattei forthcoming.

NOTES

1: CONTEXT

1 In fact, he suggests that such a way of misreading scientific texts was created by key figures in the scientific revolution itself (ET, p. xiii).
2 Perhaps Popper had Kuhn in mind when he later said he found some of the work of the Harvard graduate students 'really outstanding'.
3 My sketch of the Lowell lectures derives from the useful account in Marcum 2005, chap. 2.
4 See correspondence reproduced in Reisch 1991.
5 This article, reprinted in ET, is still perhaps the best short summary of Kuhn's early views.

2: OVERVIEW OF THEMES

1 Kuhn would not, I suspect, have described SSR as 'a polished system of philosophy' (Hacking 1981, p. 5).

3: READING *THE STRUCTURE OF SCIENTIFIC REVOLUTIONS*

1 Kuhn had attended Bridgman's lectures on thermodynamics in graduate school, and may also have attended an undergraduate course in the same subject taught by Frank (RSS, pp. 267–8). In addition, he at least knew *of* Hans Reichenbach's *Experience and Prediction* (Chicago: University of Chicago Press, 1938), since it was there that he encountered a reference to *Genesis and Development of a Scientific Fact*, by the Polish-born bacteriologist and philosopher of science Ludwik Fleck, which Kuhn found congenial.
2 Carnap, for example, in his correspondence with Kuhn, professed to have only fragmentary knowledge of the history of science (see Reisch 1991, p. 266).
3 From Kuhn's 1953 application for a Guggenheim Memorial Fellowship, quoted in Cedarbaum 1983, p. 178.

4 See, for example, Haller 1992. Positivism also had its *origin* in the work of Auguste Comte, which was both historical and historicist.

5 The sociologist of science Robert K. Merton also occasionally used similar phrases. Struan Jacobs shows that Fleck's term 'thought collective', which seems to have repelled Kuhn anyway, is simply too wide to be identified with Kuhn's idea of a scientific community (Jacobs 2002).

6 From his Guggenheim Memorial Fellowship application, quoted in Cedarbaum 1983, p. 177, emphasis added.

7 Lichtenberg also has a role in the development of the concept of revolutions in science. See Cohen 1985.

8 Kuhn later claimed that he wasn't aware of Lichtenberg's or Wittgenstein's uses of the concept (RSS, p. 299), but his friend and colleague Stanley Cavell recalls a conversation, from the mid- to late 1950s, in which Kuhn asked him about the latter.

9 Kuhn obviously took the issue of the ambiguity and vagueness of 'paradigm' seriously since, as he says, he made three attempts to recover the original sense of the term (ET, p. xx).

10 Kuhn's list of well-known classic science texts (p. 10) may itself be intended as a list of paradigms. But it is better conceived as a list of books in which paradigms (the exemplars or achievements in question) are *recounted*. (Although he mentions those who accepted Newton's *Principia* 'as paradigm', at the same time he implies that the paradigm in question is, more strictly speaking, something that resulted *from Principia* (p. 105).)

11 Note that this conflicts with allowing (immediately above) that there may be a single theory of the domain.

12 By Paul Feyerabend and Steve Fuller, for example.

13 Only '*normally* embodied', though. Joseph Agassi (1966, p. 351) is too strict in taking the existence of textbooks to be Kuhn's criterion demarcating science from pre-science. Despite some comments which might suggest otherwise (e.g. p. 137), Kuhn is usually clear that paradigms *precede* textbooks (e.g. in Crombie 1961, p. 386).

14 Elsewhere, Kuhn granted that Polanyi had 'provided the most extensive and developed discussion I know of the aspect of science which led me to my apparently strange usage [of the term 'paradigm']' (in Crombie 1961, p. 392). See Jacobs' papers on Kuhn's debts to Polanyi, listed in the bibliography. Kuhn knew of Polanyi's work from having read two of his books while teaching on Conant's course. His debts to Polanyi are several, and as important as his debts to Koyré, Wittgenstein, Quine, Fleck, or any of the other figures whose work he more fully credits.

15 See Alexander Bird's discussion of skills, intuition and connoisseurship (Bird 2000, pp. 73ff.). His explanation of these in terms of neural networks is undoubtedly close to what Kuhn had in mind.

16 Fleck, in fact, had already criticized this feature of conventionalism (Fleck 1935, pp. 8–9).

17 Kuhn may have got the concept of an anomaly either from Polanyi's *Personal Knowledge* (1958), or from Butterfield's *The Origins of Modern Science* (1949).

18 As Feyerabend, who didn't attach much importance to any alleged difference between theories and paradigms, was fond of saying: theories are *born* refuted!

19 Polanyi's use of Gestalt psychology in Chapter 4 of *Personal Knowledge* may have influenced Kuhn's, and the notion that researchers operating very different frameworks 'live in different worlds' occurs in Chapter 6 of that book (p. 151).

20 Cavell was surely influential here, since he mentions having had discussions with Kuhn about 'whether the time has come to drag free of the philosophical tradition established in response to, and as part of, the "scientific revolution" of the sixteenth and seventeenth centuries' (Cavell 1969, p. 42, n. 38).

21 Just as with his idea that there are paradigms in philosophy, one might worry that taking this seriously threatens to erode the *special* nature of scientific paradigms, which Kuhn elsewhere insists on.

22 As Stanley Cavell suggested to me, in correspondence.

23 He denies that 'seeing as' is part of '*Wahrnemung*', here probably better translated 'information pick-up', rather than 'perception' (as Anscombe's translation has it). I'm grateful to my colleague Severin Schroeder for this observation.

24 On the importance of such *similarity relations* to Kuhn's project, see especially Hoyningen-Huene 1993, chap. 3, and Bird 2000, *passim*.

25 Although the term is also sometimes used in an 'inverted commas' way, where it's obvious that what one 'sees' (e.g. a pink elephant) is *not* there to be seen. Bird (ibid., chap. 4) is quite right to criticize Kuhn's 'internalism' about perception, the idea that one is an unimpeachable authority about what one is perceiving.

26 Here I disagree with Hoyningen-Huene (1993, p. 40).

27 This might have been gleaned from his Lowell lectures, where he talked about 'perceptual' and 'behavioural' worlds.

28 However, the declaration that paradigms constitute not only science but also nature (p. 110), and his occasional use of 'nature' and 'world' as synonyms (p. 173), show that Kuhn wasn't always careful about sticking to his own terminology.

29 Winch connects this with the very same section of Wittgenstein's *Philosophical Investigations* that Kuhn was most exercised by, but there's surely also a connection here with the *Tractatus Logico-Philosophicus*' comment on solipsism that the world is *my* world (Wittgenstein 1961, proposition 5.62) *and* with the idea that solipsism, when strictly thought out, 'coincides with pure realism' (proposition 5.64).

30 In his book on moral philosophy (Arrington 1989), chap. 6.

31 Conceptual schemes are like toolkits, whereas claims are like things done with particular tools from such a kit. A paradigm, because it involves a conceptual scheme, is indeed, as Kuhn says, 'what you use when the theory isn't there' (RSS, p. 300).

32 Bird (2000, p. 130) indicates a further parallel: Kuhn's picture of the development of science is comparable to Hegel's idea of 'dialectical' development.

33　That Kuhn *did* take this seriously is evident from his reaction to the title of Fleck's *Genesis and Development of a Scientific Fact*, 'These things . . . may have an *Entstehung* [origin], but they are not supposed to have an *Entwicklung* [development]' (RSS, p. 283). What Kuhn says suggests that the title of Fleck's book attracted him because he had already had the idea that facts do indeed have an *Entwicklung*.

34　In suggesting that by changing paradigms scientists 'learn to see science and the world differently' (p. 144), Kuhn seems to have moved from the exemplar sense of 'paradigm' to the disciplinary matrix sense. (Changes of exemplar being too frequent to constitute 'world-changes'.)

35　For an important challenge to this principle, see Hull, Tessner and Diamond 1978.

36　Hacking (1983, p. 73) treats this problem as 'shallow', but doesn't say why.

37　This process continued in Kuhn's later work, where he introduced the notion of a *lexicon*, and construed incommensurability as a relation between different lexicons. See, in particular, the articles now reprinted as Chapters 3 and 4 of RSS.

38　Kuhn pursued the idea that incommensurability is local, rather than global, and distinguished between translation and *interpretation*, in his 1982 paper 'Commensurability, Comparability, Communicability', now reprinted in RSS.

39　From Cavell's notes on an early meeting with Kuhn.

40　Stephen Toulmin, David Hull and Bas van Fraassen are examples.

41　On this issue, see further Preston 2003.

BIBLIOGRAPHY

Agassi, J. (1966) 'Review of *The Structure of Scientific Revolutions*,' *Journal of the History of Philosophy*, **4**, 351–4.

Andersen, H. (2001) *On Kuhn*. Belmont, CA: Wadsworth.

Andresen, J. (1999) 'Crisis and Kuhn', *Isis*, **90** (Supplement), S43–67.

Arrington, R. L. (1989) *Rationalism, Realism and Relativism: Perspectives in Contemporary Moral Philosophy*. Ithaca: Cornell University Press.

Axtell, G. S. (1993) 'In the Tracks of the Historicist Movement: Re-assessing the Carnap-Kuhn Connection', *Studies in History and Philosophy of Science*, **24**, 119–46.

Bachelard, G. (1927) *Étude sur l'evolution d'un problème de physique: la propagation thermique dans les solides*. Paris: J.Vrin.

Biagioli, M. (1990) 'The Anthropology of Incommensurability', *Studies in History and Philosophy of Science*, **21**, 183–209.

Bird, A. (2000) *Thomas Kuhn*. Chesham, Bucks: Acumen.

——(2002) 'Kuhn's Wrong Turning', *Studies in History and Philosophy of Science*, **33**, 443–63.

——(2004) 'Kuhn, Naturalism, and the Positivist Legacy', *Studies in History and Philosophy of Science*, **35**, 337–56.

Bohm, D. (1964) Review of *The Structure of Scientific Revolutions*, *The Philosophical Quarterly*, **14**, 377–9.

Brown, H. I. (1979) *Perception, Theory and Commitment: The New Philosophy of Science*. Chicago: University of Chicago Press.

——(1983) 'Incommensurability', *Inquiry*, **26**, 3–29.

Bruner, J. S. and Postman, L. (1949) 'On the Perception of Incongruity: A Paradigm', *Journal of Personality*, **18**, 206–23.

Buchwald, J. Z. and Smith, G. E. (1997) 'Thomas Kuhn, 1922–1996', *Philosophy of Science*, **64**, 361–76.

——(2001) 'Kuhn and Incommensurability', *Perspectives on Science*, **9**, 463–98.

Burian, R. M. (1984) 'Scientific Realism and Incommensurability: Some Criticisms of Kuhn and Feyerabend', in R. S. Cohen and M. W. Wartofsky (eds), *Methodology, Metaphysics and the History of Science*, (pp. 1–31). Dordrecht: D. Reidel.

Butterfield, H. (1931) *The Whig Interpretation of History*. London: Bell & Sons. (Reprinted Penguin, Harmondsworth, UK, 1973.)
——(1949) *The Origins of Modern Science, 1300–1800*. London: Bell & Hyman.
Cavell, S. (1969) *Must We Mean What We Say?* New York: Scribner's.
Cedarbaum, D. G. (1983) 'Paradigms', *Studies in History and Philosophy of Science*, **14**, 173–213.
Chen, X. (1997) 'Thomas Kuhn's Latest Notion of Incommensurability', *Journal for General Philosophy of Science*, **28**, 257–73.
Cohen, I. B. (1985) *Revolution in Science*. Cambridge, MA: Harvard University Press.
Collier, J. (1984) 'Pragmatic Incommensurability', in P. D. Asquith and P. Kitcher (eds), *PSA 1984, Volume 1*, (pp. 146–53). East Lansing, MI: Philosophy of Science Association.
Crombie, A. C. (ed.) (1963) *Scientific Change*. London: Heinemann, 1963.
Davidson, D. (1984 [1974]) 'On the Very Idea of a Conceptual Scheme', [1974], reprinted in his *Inquiries into Truth and Interpretation*, (pp. 183–98). Oxford: Clarendon Press.
Davidson, K. (forthcoming) *The Death of Truth: Thomas S. Kuhn and the Evolution of Ideas*. Oxford: Oxford University Press.
Devitt, M. (1979) 'Against Incommensurability', *Australasian Journal of Philosophy*, **57**, 29–50.
Donovan, A., Laudan, L. and Laudan, R. (eds) (1988) *Scrutinizing Science: Empirical Studies of Scientific Change*. Baltimore, MD: John Hopkins University Press.
Doppelt, G. (1978) 'Kuhn's Epistemological Relativism: An Interpretation and Defence', *Inquiry*, **21**, 33–86.
Feyerabend, P. K. (1964) Review of A. C. Crombie (ed.) *Scientific Change*, *British Journal for the Philosophy of Science*, **15**, 244–54.
——(1975) *Against Method*. London: New Left Books.
——(1978) *Science in a Free Society*. London: New Left Books.
——(1981a) *Realism, Rationalism and Scientific Method: Philosophical Papers, Volume 1*. Cambridge: Cambridge University Press.
——(1981b) *Problems of Empiricism: Philosophical Papers, Volume 2*. Cambridge: Cambridge University Press.
——(1999) *Knowledge, Science and Relativism: Philosophical Papers, Volume 3*, ed. J. M. Preston. Cambridge: Cambridge University Press.
Fleck, L. (1979) *Genesis and Development of a Scientific Fact* [1935]. Chicago: University Press.
Frank, P. (ed.) (1954) *The Validation of Scientific Theories*. Boston: Beacon Press.
Franklin, A. (1984) 'Are Paradigms Incommensurable?', *British Journal for the Philosophy of Science*, **35**, 57–60.
Friedman, M. (1999) *Reconsidering Logical Positivism*. Cambridge: Cambridge University Press.
——(2001) *Dynamics of Reason*. Stanford, CA: CSLI.
Fuller, S. (2000) *Thomas Kuhn: A Philosophical History for Our Times*. Chicago: University of Chicago Press.

Fuller, S. (2003) *Kuhn versus Popper: The Struggle for the Soul of Science*. Duxford: Icon Books.

Gattei, S. (forthcoming) *Thomas S. Kuhn's 'Linguistic Turn' and the Legacy of Logical Positivism*. Aldershot: Ashgate.

Geertz, C. (2000) 'The Legacy of Thomas Kuhn: The Right Text at the Right Time', in his *Available Light: Anthropological Reflections on Philosophical Topics*, (pp. 160–6). Princeton, NJ: Princeton University Press.

Glock, H. J. (2000) 'Imposters, Bunglers and Relativists', in S. Peters and M. Biddis (eds), *The Humanities in the New Millenium*, (pp. 267–87). Tübingen and Basel: A. Francke Verlag.

——(2008) 'Relativism, Commensurability and Translatability', in J. M. Preston (ed.), *Wittgenstein and Reason*. Oxford: Blackwell.

Grandy, R. (1983) 'Incommensurability: Kinds and Causes', *Philosophica*, **32**, 7–24.

Gutting, G. (ed.) (1980) *Paradigms and Revolutions: Appraisals and Applications of Thomas Kuhn's Philosophy of Science*. Notre Dame, IN: University Press.

Hacker, P. M. S. (1996) 'On Davidson's Idea of a Conceptual Scheme', *The Philosophical Quarterly*, **46**, 289–307.

Hacking, I. (1979) Review of T. S. Kuhn, *The Essential Tension, History and Theory*, **18**, 223–36.

——(ed.) (1981) *Scientific Revolutions*. Oxford: Oxford University Press.

——(1983) *Representing and Intervening: Introductory Topics in the Philosophy of Natural Science*. Cambridge: Cambridge University Press.

——(1985) 'Styles of Scientific Reasoning', in J. Rajchmann and C. West (eds), *Post-Analytic Philosophy*, (pp. 145–65). New York: Columbia University Press.

——(1986) 'Science Turned Upside Down' (Review of I. B. Cohen, *Revolution in Science*), *New York Review of Books*, 27 February, 21–5.

Haller, R. (1992) 'The First Vienna Circle', in T. E. Uebel (ed.), *Rediscovering the Forgotten Vienna Circle*, (pp. 95–108). (Dordrecht: Kluwer.

Hanson, N. R. (1965) 'A Note on Kuhn's Method', *Dialogue*, **4**, 371–5.

Heilbron, J. L. (1998) 'Thomas Samuel Kuhn', *Isis*, **89**, 505–15.

Hesse, M. B. (1963) Review of *The Structure of Scientific Revolutions*, *Isis*, **54**, 286–7.

Horwich, P. (ed.) (1993) *World Changes: Thomas Kuhn and the Nature of Science*. Cambridge, MA: MIT Press.

Hoyningen-Huene, P. (1993) *Reconstructing Scientific Revolutions: Thomas S. Kuhn's Philosophy of Science*. Chicago: University of Chicago Press.

——(1995) 'Two Letters of Paul Feyerabend to Thomas S. Kuhn on a Draft of *The Structure of Scientific Revolutions*', *Studies in History and Philosophy of Science*, **26**, 353–87.

——(2000) 'Paul Feyerabend and Thomas Kuhn', in J. M. Preston, G. Munévar and D. Lamb (eds), *The Worst Enemy of Science?: Essays in Memory of Paul Feyerabend*, (pp. 102–14). New York: Oxford University Press.

——(2006) 'More Letters by Paul Feyerabend to Thomas S. Kuhn on *Proto-Structure*', *Studies in History and Philosophy of Science*, **37**, 610–32.

Hoyningen, P. and Sankey, H. (eds) (2001) *Incommensurability and Related Matters*. Dordrecht: Kluwer.

Hoyningen-Huene, P., Oberheim, E. and Andersen, H. (1996) 'On Incommensurability', *Studies in History and Philosophy of Science*, **27**, 131–41.

Hull, D. L., Tessner, P. D. and Diamond, A. M. (1978) 'Planck's Principle', *Science*, **202**, 717–23.

Irzik, G. and Grunberg, T. (1995) 'Carnap and Kuhn: Arch-Enemies or Close Allies?', *British Journal for the Philosophy of Science*, **46**, 285–307.

——(1998) 'Whorfian Variations on Kantian Themes: Kuhn's Linguistic Turn', *Studies in History and Philosophy of Science*, **29**, 207–21.

Jacobs, S. (2002) 'Polanyi's Presagement of the Incommensurability Concept', *Studies in History and Philosophy of Science*, **33**, 105–20.

——(2002) 'The Genesis of "Scientific Community"', *Social Epistemology*, **16**, 157–68.

Kitcher, P. (1983) 'Implications of Incommensurability', in P. D. Asquith and T. Nickles (eds), *PSA 1982, Volume 2*, (pp. 689–703). East Lansing, MI: Philosophy of Science Association.

Kindi, V. (1994) 'Incommensurability, Incomparability, Irrationality', *Methodology and Science*, **27**, 40–55.

——(1995) 'Kuhn's *The Structure of Scientific Revolutions* Revisited', *Journal for General Philosophy of Science*, **26**, 75–92.

Koyré, A. (1954) 'Influence of Philosophic Trends on the Formulation of Scientific Theories', in Frank (1954), (pp. 177–87).

——(1978) *Galileo Studies*. Sussex: Harvester Press.

Kourany, J. A. (1979) 'The Nonhistorical Basis of Kuhn's Theory of Science', *Nature & System*, **1**, 46–59.

Kuhn, T. S. (1963) 'The Function of Dogma in Scientific Research', in A. C. Crombie (ed.), *Scientific Change*, (pp. 347–95, plus discussion comments on pp. 386–95). London: Heinemann.

——(1970) 'Alexandre Koyré and the History of Science: On an Intellectual Revolution', *Encounter*, **34**, 67–70.

Lakatos, I. and Musgrave, A. E. (eds) (1970) *Criticism and the Growth of Knowledge*. Cambridge: Cambridge University Press.

Larvor, B. (2003) 'Why did Kuhn's *Structure of Scientific Revolutions* Cause a Fuss?', *Studies in History and Philosophy of Science*, **34**, 369–90.

Lovejoy, A. O. (1936) *The Great Chain of Being*. Cambridge, MA: Harvard University Press.

McGrew, T. (1994) 'Scientific Progress, Relativism, and Self-Refutation', *The Electronic Journal of Analytic Philosophy*, **2**.

MacIntyre, A. (1977) 'Epistemological Crises, Dramatic Narrative and the Philosophy of Science', *The Monist*, **60**, 453–72. (Reprinted in Gutting 1980.)

Malone, M. E. (1993) 'Kuhn Reconstructed: Incommensurability without Relativism', *Studies in History and Philosophy of Science*, **24**, 69–93.

Marcum, J. A. (2005) *Thomas Kuhn's Revolution: An Historical Philosophy of Science*. London: Continuum.

Meiland, J. W. (1974) 'Kuhn, Scheffler, and Objectivity in Science', *Philosophy of Science*, **41**, 179–87.

Moberg, D. W. (1979) 'Are there Rival, Incommensurable Theories?', *Philosophy of Science*, **46**, 244–62.

Musgrave, A. E. (1979) 'How to Avoid Incommensurability', *Acta Philosophica Fennica*, **30**, 337–46.

Nagel, E. (1961) *The Structure of Science*. London: Routledge & Kegan Paul.

Nickles, T. (ed.) (2002) *Thomas Kuhn*. Cambridge: Cambridge University Press.

Pinto de Oliveira, J. C. (2007) 'Carnal, Kuhn, and Revisionism: on the publication of *Structure* in *Encyclopedia*', *Journal for General Philosophy of Science*, 38, 147–57.

Polanyi, M. (1958) *Personal Knowledge: Towards a Post-Critical Philosophy*. Chicago: University of Chicago Press.

Popper, K. R. (1959) *The Logic of Scientific Discovery*. London: Hutchinson.

——(1974) 'Replies to my Critics', in P. A. Schilpp (ed.), *The Philosophy of Karl Popper*, (pp. 961–1197). LaSalle, IL: Open Court.

Preston, J. M. (1997) *Feyerabend: Philosophy, Science and Society*. Cambridge: Polity Press.

——(2003) 'Kuhn, Instrumentalism, and the Progress of Science', *Social Epistemology*, **17**, 259–65.

——(2004) 'Bird, Kuhn and Positivism', *Studies in History and Philosophy of Science*, **35**, 327–35.

Reisch, G. A. (1991) 'Did Kuhn Kill Logical Empiricism?', *Philosophy of Science*, **58**, 264–77.

Robinson, G. (1996) 'On Misunderstanding Science', *International Journal of Philosophical Studies*, **4**, 110–27.

Rorty, R. (2000) 'Kuhn', in W. H. Newton-Smith (ed.), *A Companion to the Philosophy of Science*, (pp. 203–6). Oxford: Blackwell.

Sankey, H. (1994) *The Incommensurability Thesis*. Aldershot: Avebury Press.

——(1997) *Rationality, Relativism and Incommensurability*. Aldershot: Avebury Press.

Sarton, G. (1936) *The Study of the History of Science*. Cambridge, MA: Harvard University Press.

Scheffler, I. (1967) *Science and Subjectivity*, (chap. 4). New York: Bobbs-Merrill.

——(1972) 'Vision and Revolution: A Postscript on Kuhn', *Philosophy of Science*, **39**, 366–74. (Reprinted in Scheffler 1982.)

——(1982) *Science and Subjectivity*, (2nd edn). Indianapolis: Hackett.

Shapere, D. (1964) Review of T.S.Kuhn, *The Structure of Scientific Revolutions*, *The Philosophical Review*, **73**, 383–94.

——(1971) 'The Paradigm Concept', *Science*, **172**, 706–9.

——(1981) 'Meaning and Scientific Change', in R. G. Colodny (ed.), *Mind and Cosmos: Essays in Contemporary Science and Philosophy*, (pp. 41–85). Pittsburgh: University of Pittsburgh Press. (Partly reprinted in Hacking 1981.)

Sharrock, W. and Read, R. (2002) *Kuhn: Philosopher of Scientific Revolution*. Cambridge: Polity Press.

Short, T. L. (1980) 'Peirce and the Incommensurability of Theories', *The Monist*, **63**, 316–28.

Siegel, H. (1987) *Relativism Refuted: A Critique of Contemporary Epistemological Relativism*. Dordrecht: D. Reidel.

Sigurdsonn, S. (1990) 'The Nature of Scientific Knowledge: An Interview with Thomas Kuhn', *Harvard Science Review*, Winter, 18–25.

Sokal, A. and Bricmont, J. (1998) *Intellectual Impostures*. London: Profile Books.

Stove, D. C. (1982) *Popper and After: Four Modern Irrationalists*. Oxford: Pergamon Press.

Suppe, F. (ed.) (1977) *The Structure of Scientific Theories*, (2nd edn). Urbana: University of Illinois Press.

Szumilewicz, I. (1977) 'Incommensurability and the Rationality of the Development of Science', *British Journal for the Philosophy of Science*, **28**, 345–50.

Thagard, P. (1992) *Conceptual Revolutions*. New Jersey: Princeton University Press.

Toulmin, S. E. (1972) *Human Understanding, Volume 1: The Collective Use and Evolution of Concepts*. Oxford: Clarendon Press.

von Dietze, E. (2001) *Paradigms Explained: Rethinking Thomas Kuhn's Philosophy of Science*. Westport, CT: Praeger.

Whewell, W. (1984) *Selected Writings on the History of Science*, ed. Y. Elkana. Chicago: University of Chicago Press.

Winch, P. (1958) *The Idea of a Social Science, and its Relation to Philosophy*. London: Routledge.

Wisdom, J. O. (1974) 'The Incommensurability Thesis', *Philosophical Studies*, **25**, 299–301.

Wittgenstein, L. (1961) *Tractatus Logico-Philosophicus*, trans D. F. Pears and B. F. McGuinness. London: Routledge & Kegan Paul.

——(1958) *Philosophical Investigations*, (2nd edn), trans G. E. M. Anscombe. Oxford: Blackwell.

Wong, D. B. (1989) 'Three Kinds of Incommensurability', in M. Krausz (ed.), *Relativism: Interpretation and Confrontation*, (pp. 140–58). Notre Dame, IN: University of Notre Dame Press.

INDEX

aesthetic factors 86, 88
anomalies 5, 11, 30, 40–51, 55,
    113n
Archimedes 26
Aristotle 2, 24, 26, 65, 66
Arrington, Robert 76, 78
aspect-perception 66–72
avowals 68, 70, 70

Bachelard, Gaston 4
Bacon, Sir Francis 25
Bird, Alexander 99–100, 107, 109,
    113n, 114n
Black, Joseph 26
Boerhaave, Herman 26
Bohm, David 36, 104
Boyle, Robert 26, 83
Bridgman, Percy W. 14, 112n
Bruner, Jerome 43, 68, 69
Butterfield, Herbert 3, 81, 110,
    113n

caloric theory of heat 59
Carnap, Rudolf 6, 14, 16–17, 18,
    110, 112n
Cavell, Stanley 6, 7, 24–5, 97, 113n,
    114n, 115n
Cohen, I. Bernard 3
community, communities, scientific
    15, 19, 36, 38–9, 45, 47, 60, 61,
    85, 88, 91–2, 95–8, 102, 103,
    113n
Comte, Auguste 3, 113n

Conant, James Bryant 2, 3, 5, 7, 9,
    53, 113n
conceptual scheme(s) 11, 45, 73,
    76–8, 101–2, 114n
    see also relativism, conceptual
confirmation 79, 84
conventionalism 17, 36, 93, 113n
conversion 85, 93, 94
Copernicus, Nicolaus 59
corpuscular theory 24, 39, 59
crises 5, 11, 17, 46–51, 55
critical rationalism 12, 15, 16, 21
cumulativism, cumulative progress
    12, 17, 18, 19, 40, 52, 53, 57,
    62, 80, 81, 82, 94, 95, 98, 111

Dalton, John 59
Darwin, Charles 59
Darwinian history 54, 99
Davidson, Donald 110
Descartes, René 58
disciplinary matrix 23, 24, 28, 39,
    45, 56, 59, 60, 62, 63, 87, 89,
    101–2, 115n
discovery 40–5, 50
Duhem, Pierre 36, 54,

Eddington, Sir Arthur 14
education, scientific 63, 78, 82, 109
Einstein, Albert 55–6, 57, 59, 89
empiricism, Logical Empiricism 6,
    12, 15, 17, 18, 21, 31, 36, 53,
    54, 55, 75, 84, 89, 90, 106, 110